Managerial effectiveness in fisheries and aquaculture

Managerial effectiveness in fisheries and aquaculture

Ian Chaston BSc, PhD, MBA

Fishing News Books Ltd
Farnham, Surrey, England

© Ian Chaston 1988

British Library CIP Data

Chaston, Ian
  Managerial effectiveness in fisheries and
  aquaculture
  1. Fishing industries. Management
  2. Aquaculture industries. Management
  I. Title
  639′.2′068

ISBN 0-85238-157-3

Published by
Fishing News Books Ltd
1 Long Garden Walk
Farnham, Surrey, England

Typeset by
Mathematical Composition Setters Ltd
Salisbury, Wiltshire

Printed in Great Britain by
Henry Ling Ltd
The Dorset Press, Dorchester

Contents

Illustrations and tables

Preface

Low profitability and even business failure are common occurrences in the fisheries and aquaculture industry. Companies frequently explain their poor performance as being caused by factors outside their control such as over-fishing, rising energy costs, currency fluctuations, unforeseen technological obstacles and trade barriers. In many cases, however, the real explanation is that the companies are staffed by executives who have inadequate managerial skills.

Executives in fisheries and aquaculture are usually appointed to a management role because they exhibited excellent functional skills in a specialist area (*eg* engineering or accountancy). Unfortunately the appointment will involve being responsible for handling subordinates. These latter individuals, as members of the human race, can offer loyalty, support and co-operation. Just as possibly, however, the manager will find subordinates can be irritable, irresponsible and obstructive. Hence, to be effective, all managers have to develop the inter-personal skills of being able to work with and through subordinates to optimise organisational performance.

The purpose of the text is to introduce selected aspects of the behavioural skills that are required of individuals if they are to be effective managers. The topics covered are analysis of the management role, communication, motivation, performance appraisal, planning, decision making and control. These concepts are presented by illustrating their application to problems encountered by managers in the fisheries and aquaculture industry. Although the examples used in the text are based upon situations known to the author, the reader is cautioned that a number of the cases have been simplified to highlight a specific issue. In real life, managers will find that problems, especially those involving subordinates, are often more complex because of the interaction between the numerous variables which exist in the corporate environment.

The text is designed for use by practising managers and public sector advisors in the industry, as well as providing the basis for a college level course on the management role in fisheries and aquaculture.

Recognition is due to Beryl, Roy R and Simon for providing the idea for the text and to Roy N for a contrasting opinion based on a more numeric philosophy.

Ian Chaston

1 What do managers do?

Case: Scotia Fisheries Ltd

John Bliss is the plant manager of Scotia Fisheries processing plant which produces frozen and fresh groundfish fillets. The plant is located on the coast, approximately 400 miles away from the company Head Office.

A typical day for Mr Bliss starts at 6.30 a.m. when he arrives at the plant in time to talk to employees at the end of the night shift, and to review progress of the quality control clean-up crew preparing for the 8.00 a.m. day shift. He allocates an hour between 7.30 – 8.30 a.m. to reading the mail, examining production reports from the previous day and reviewing the current day's production schedules. Mr Bliss has a reputation for always being out in the plant for most of the day. Hence employees at other Scotia Fisheries locations (*eg* the accountants and sales staff at Head Office), know that early morning is a good time to phone him.

From 8.30 a.m. to about 11.30 a.m., Mr Bliss spends time visiting various areas of the plant operation (*eg* fish unloading/grading; filleting and skinning; freezing and packing) to discuss any problems which may have arisen at the beginning of the new shift.

Mr Bliss usually has lunch at his desk, but today the plant is being visited by a Fisheries Department technician who wants to show some new bacterial analysis techniques to the quality control department. As Mr Bliss feels it is important to keep in touch with Government departments, he invites the technician and the head of the quality control department to have lunch with him at a local hotel. On the way back from lunch, Mr Bliss stopped off to talk to the mayor about the forthcoming town barbecue which Scotia Fisheries, as the largest employer in the area, always sponsors. As a result he arrived back at the plant some 30 minutes late for the beginning of the weekly meeting of department heads. They decided to start without him. Believing he should not ask his staff to go back over topics already discussed, he proposed to catch up on these matters the next day with relevant department heads.

The meeting was over by 2.30 p.m., and for the next 30 minutes Mr Bliss dealt with the more urgent telephone messages which had accumulated. He had intended to spend the next hour with his secretary going through his correspondence. After a few minutes he saw the head of maintenance rush past his office, and following him out to the processing area, found that a conveyor belt had broken. For the next hour, Mr Bliss and the line supervisors were involved in revising the flow of products through the packing area in order to avoid a build up of fillets. At 4.00 p.m. he went over to the filleting area to talk to the maintenance crew about his new idea to adjust the cutting angle on the skinning machines. He spent the next hour working with the crew on a series of tests to evaluate the effect of his proposed adjustments on fillet yields.

At 5.00 p.m. Mr Bliss went home, taking with him the correspondence he had not been able to handle during the day. After dinner he worked through various reports and letters, dictating answers into his portable machine or making notes for discussion with his staff. Before going to bed, he went down to the plant for a few minutes to make sure the night shift were not encountering any problems with the repaired conveyor belt.

Question: Do you judge Mr Bliss to be an effective manager?

In considering this issue, it might be worth noting the comments of other personnel within Scotia Fisheries.

The Production Director: 'John's one hell of a good man to have around. He runs that plant like a well-oiled machine. The only real complaint I ever have is not getting his production reports and financial summaries on time. All in all, I do not look forward to his retirement, I just do not know who we have at the plant that will be capable of replacing him'.

A Department Head: 'Mr Bliss always puts in 110%, he's everywhere, every day, checking, suggesting and helping. Sometimes I wonder whether he really needs the rest of us'.

A Production Operative: 'Mr Bliss is a good boss. His door's always open and if we do not like a decision by a department head or line supervisor, he will usually give us a fair hearing'.

The manager's tasks – fact and fiction

Until the early 1970s, the accepted definition of the key tasks of a manager was that the individual was responsible for planning, organising, co-ordinating and controlling. In consequence, for almost half a century, the field of management studies has centred on methods to improve the provision of information to assist managers to make better decisions. Concurrently, training systems were designed to improve the ability of managers to execute the four key tasks.

Over the last two decades new understanding of the role of managers has been provided by behavioural scientists. These researchers observed the activities of managers in day-to-day situations across a broad range of organisations. The picture which has emerged is actually rather obvious to anybody who has spent time in a manager's office – either behind, or in front of the desk. Yet this research has helped throw into question many of the classic theories on management and led to a much better understanding of how to improve managerial performance in both the private and public sector.

The research showed that the manager is not a reflective, systematic planner. The studies soon proved that managers have to work at an unrelenting pace, moving from one situation to the next in response to the needs of the moment. Under these circumstances, there is rarely any opportunity to carefully plan, adhere to a specific schedule, and pause to regularly analyse the implications of even major issues. When the manager is forced to plan, he usually does this through an intuitive, internalised process based upon experience of previous situations.

Post-war management theory has stressed the need for the manager to delegate standard tasks in order to release him to orchestrate over-all corporate programmes. In real life, however, managers will be found performing regular tasks such as negotiating with suppliers and processing information, which in theory could be handled by subordinates. The explanation for this is simple. Most organisations cannot afford to hire a workforce large enough to allow managers, including the company president, to be free from everyday activities.

The advent of the computer has been hailed as the breakthrough to ensure managers can be kept informed by data bases which describe performance across the company. It is now apparent, however, that managers avoid hard information. Instead they prefer direct contact with other individuals using the verbal media of informal discussions, telephone calls and meetings to collect information. Furthermore most managers cherish 'soft' information, especially gossip, hearsay and speculation. Carefully presented statements about plans and actual performance versus budget, are not received with anything like the enthusiasm as the latest office gossip about next year's salary increases.

This preference for verbal media has two important implications. Firstly, verbal information is stored in our minds, not in the files or data banks of the organisation. By definition, therefore, the records within any company must be incomplete. Secondly, this reliance on verbal data helps to explain the reluctance to delegate. As information 'on file' is incomplete, the manager cannot hand over a complete summary to the subordinate with instructions on what to do next. It will be necessary to brief the individual on the various issues not completely covered by the company records. This can take time, and thus the typical reaction of the manager is that it is 'easier to do it myself'.

Upon examining the variance between the theory and realities of management, it becomes apparent that the manager's job is immensely complicated. The individual is overburdened with responsibilities, yet not easily able to delegate anything other than the simplest of tasks. As a result, he is driven to overwork and forced to handle many issues superficially. Furthermore the nature of modern society is increasing the pressures now being placed on managers. Subordinates now expect to question decisions, and a growing number of outside bodies (eg governmental, consumer groups and customers) demand access in order to influence the actions of an organisation. Hence to develop any techniques to help a manager be more effective, we have to discard previous theories and begin to define what the role is really about.

Elements of the managerial function

The leading researcher in management behaviour, H. Mintzberg, has proposed ten roles, or organised sets of behaviour. Three roles – figurehead, leader and liaison – involve inter-personal activities. Another three – monitor, disseminator and spokesman – are concerned with informational processes. The final four – entrepreneur, disturbance handler, resource allocator and negotiator – are all decisional in nature (Fig 1.1).

The figurehead role is created by virtue of the manager being the head of an organisational unit (eg the production supervisor handing out prizes at the factory dance; the company president greeting a politician invited to open a new processing facility). Because the manager is placed in charge of a group of employees, he will be responsible for ensuring that they perform

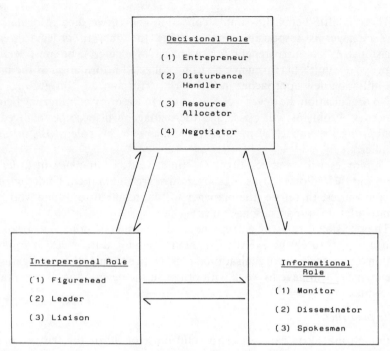

Fig 1.1 Managerial roles

their allocated tasks. To achieve this objective, the manager will have to be a leader. This involves both the direct actions associated with controlling employee behaviour and the more intangible process of motivating them through encouragement to perform effectively.

Organisational units cannot operate in isolation from other groups within the company. To ensure productive interaction between units inside and outside the organisation, the unit manager must act as a link or 'liaison point' for the employees who report to him.

The manager's liaison role consists of monitoring all channels of communication within the organisation in order to collect information. These data are then assimilated prior to some being shared with others. In the disseminator role, the manager passes on some of the collected information to subordinates who, in many cases, have no direct access to it. The manager's own unit will in turn generate information which has been passed on to others. In this case the manager acts as the spokesman, communicating with others in the organisation and in the external environment in which the company operates.

There is little benefit in simply collecting and forwarding information. The manager is responsible for applying the data in order to reach decisions. As an entrepreneur (or 'manager of change'), he must seek to improve the unit's performance by using collected information as the basis for initiating new approaches in the work performance of his group.

No organisation, however well run, can plan for every contingency before it occurs. Problems will arise, and the manager will become involved in minimising the impact of unforeseen problems in his role of disturbance handler.

Employees will need resources (*eg* time, staff, equipment) in order to carry out their allotted tasks. Resources are rarely unlimited. Under normal circumstances, therefore, the manager will be required to decide who gets what within his organisational unit.

The decisional roles of entrepreneur, disturbance handler and resource allocator can rarely be executed without resolving differences of opinion within and outside the organisational unit. Hence to fulfil these obligations, the manager will have to work with others in the negotiation of acceptable solutions.

Role priorities

Although managers can expect to fulfil most or all of the ten roles, the importance of the various activities will vary depending upon the position of the individual within the organisational hierarchy and the structure of the organisation itself. The head of a large diversified fish processing company could be expected to spend the greater part of his time acting in a leadership capacity. Most of the informational roles would be delegated. Nevertheless in the overall framework, he would be expected as the figurehead to be involved in the key task of acting as the spokesman for the organisation with external groups such as shareholders or financial institutions. He would also be expected to be a major decisional force overseeing those issues of corporate policy which would significantly influence the performance of the company over the long term.

A large fish processing company would probably be organised on a hierarchical basis with a number of levels of management in both the head office and at satellite operations elsewhere in the world. A manager in this type of company in, for example, the credit control department, would have roles of differing importance compared to those of the company's president. The credit control department, being a lower level of management, would have its personnel spending most of their time on the informational duties associated with the monitoring of performance and the dissemination of results. Decisional roles would mainly focus on disturbance handling and resource allocation.

In small companies (*eg* a trout farm), the organisation is not of a size sufficient to require numerous levels of management. Typically, the head of the operation (who is often the major shareholder) will not only have to execute the role equivalent to that of the president in a large company, but also spends more time on informational tasks and allocating resources. If this type of company is successful and expands, it is not uncommon to find that the owner/manager finds it difficult to relinquish these responsibilities to the newly-hired lower management. Under these circumstances, potential growth rates are at risk unless the owner/manager can be persuaded to delegate day-to-day matters and to dedicate more time to leadership/figurehead/liaison roles.

Management effectiveness

In order to be effective, a manager must (a) understand the relative importance of the ten roles of management for the specific position to which he has been appointed, and (b) critically determine if role priorities are adhered to in the day-to-day execution of management tasks.

To establish the relative priorities for any given management position, the manager can either refer to the job specification, if this is documented within the organisation, or discuss the subject with a superior. He can then review his activities during a typical day to see if role priorities are being correctly met.

This type of analysis frequently reveals that the manager is placing incorrect priorities on the ten roles. The usual reason for this is that we avoid those tasks in which we lack confidence and overemphasise those we find easier or most agreeable to complete.

If a significant variance between specified and actual roles is identified, it is worth examining the individual's approach to management to see if there is an actual avoidance of certain responsibilities. Where this is found to be the case, the problem can usually be overcome by (a) making the individual aware of the variance, and (b) designing a training programme to improve the skills (and consequently the confidence) of the manager in the under-applied role areas.

Scotia Fisheries – management roles

The Production Director is a little concerned about how John Bliss approaches his job as Plant Manager. In addition, Scotia is in the process of buying two small processing operations from another company. If this acquisition occurs, the Production Director would like to appoint John as Regional Plant Operations Manager responsible for all fish processing plants. On the basis of performance at the existing plant, he is concerned about John's ability to handle additional responsibilities.

He decided to approach the problem by defining the required role priorities of a Plant Manager. The next time he was at the plant, he had an informal conversation with John over the concept of role priorities and

Table 1.1   Comparison of the rank order of importance of the ten roles which constitute the responsibilities of the Plant Manager at Scotia Fisheries

Managerial roles	Production Director's desired ranking for a plant manager[*]	John Bliss self ranking of plant manager role[*]	Variance in the ranking
FIGUREHEAD- formal representative of group on 'external occasions'	2	3	-1
GROUP LEADER- responsible for managing group and getting desired actions	1	1	0
LIAISON- communicating with other groups on behalf of group	6	4	+2
MONITOR- acquiring information from group and from other groups inside/ outside organisation	3	7	-4
DISSEMINATOR- communicating information from within group and throughout organisation	5	10	-5
SPOKESMAN- represent group to others both inside and outside the organisation	4	2	+2
ENTREPRENEUR- assisting and developing new ideas to solve problems or exploit opportunities	9	5	+4
DISTURBANCE HANDLER- coping with and resolving conflicts and discipline problems	10	6	+4
RESOURCE ALLOCATOR- deciding what work is done by individuals	7	9	-2
NEGOTIATOR- bargaining with others both inside and outside the group	8	8	0

([*]Ranking 1 to 10 in descending order of importance, ie 1 is the most important role and 10 the least important)

asked him to complete a ranking on how he approached his job. The results of this analysis are shown in *Table 1.1.*

Examination of the variance data in *Table 1.1* shows that John Bliss is under-emphasising the disseminator/resource allocator role because he is allocating an excessive priority to liaison/entrepreneur/spokesman roles. This conclusion is supported by the comments of the department head that 'he's everywhere, every day.....wonder whether he really needs the rest of us'.

Compared to many people, John Bliss is a good manager; but there is room for improvement. If he is to give more priority to the monitor/disseminator/resource allocator roles, then quite obviously he must learn the art of delegation. This will require him to develop a greater level of faith in his subordinates to carry out their responsibilities, and may even involve management training programmes for both them and himself. Only then can the lower level managers in the plant begin to operate with a greater degree of personal responsibility. Certainly unless John Bliss has the potential for such improvement, he would not be a suitable candidate for promotion to the post of Regional Plant Operations Manager.

2 Communication

Case: Blue Fin Ltd

Blue Fin is a fish and seafood wholesaler supplying the catering trade through five regional storage depots. In reaction to a recent downturn in sales, the Sales Director Mr John Dennis has arranged to have tray packing systems installed in each depot. The objective is to use tray packed products as the basis for entering the retail sector, supplying small retail outlets.

A month after launching the new product range, Mr Dennis was surprised to learn that sales had been virtually zero. Upon contacting the depot managers, he found that although many retailers wished to stock the products, any attempts to generate sales were frustrated by the slowness of the credit control department at Head Office to approve the new customers as acceptable credit risks. Mr Dennis was going to be out of the office attending sales meetings for the next few days, so he dictated the following memo to his counterpart in the accounting department:

Interoffice memo

To: Mr A Balfour   From: Mr J Dennis
Accounting Director   Sales Director

Given that you have been emphasising the poor sales results over recent months at every management meeting, I would have thought you recognised the importance of the new tray pack products. It would therefore be an excellent idea if you did something about your credit department operation as soon as possible. Unless my van salesmen start getting some co-operation over a more rapid approval of credit for new customers, I will be forced to submit a revised sales forecast to the Board.

Rest assured that if I do issue a new forecast, everybody is going to be made fully aware of the real cause of my problem.

Mr Balfour received the memo on Monday morning, and having found that he could not speak to Mr Dennis until Wednesday, he dictated the following response:

Interoffice memo

To: Mr J Dennis   From: Mr A Balfour
Sales Director   Accounting Director

Had you paid attention to my last report on the average age of the company's accounts receivables, you would fully understand the priority of our credit department. Over the last six months, the average collection period on outstanding accounts has risen from 45 to 62 days. There is, therefore, little point in adding any new customers if their inability to pay their bills only compounds our cash flow situation.

My staff are quite correctly placing the majority of their efforts in speeding up the collection of outstanding balances among existing customers. Furthermore when reviewing requests to approve new customers, they are taking great care not to approve any who have a reputation for being 'slow payers'.

Hence I refuse to accept that my department is acting in any other way than that designed to ensure the company is effectively utilising every available asset.

It is apparent that the above interchange is unlikely to influence future sales performance or alter the credit department operating practices. Quite simply, Blue Fin is experiencing a failure in communication between two senior managers. Unfortunately this is not uncommon in many organisations.

Time spent communicating

Research such as that done by R Stewart has shown that managers spend about 70–80% of their time receiving or transmitting information. This occurs through the media of meetings, face to face dialogue, by telephone, reports, memoranda and letters. The information transfer process can be both incoming (initiated by others) or outgoing (initiated by the manager). A typical balance of information flow for a manager could be as follows:

Other party involved	Incoming	Outgoing
Superiors	5%	4%
People outside organisation	6%	8%
Colleagues inside organisation	11%	14%

Subordinates	15%	16%

Balance of activities not
involving information transfer 21%

Given the amount of time involved in handling information, it is imperative that managers communicate effectively. Communication problems will occur, as demonstrated by the Blue Fin case, because people say and hear things differently. In order to fully understand the skills involved in effective communication, therefore, it is necessary to examine the potential barriers to the transfer of information.

The communications process

A simple model of the communications process can be constructed by observing a warehouse supervisor instructing a fork lift truck operator where he would like a delivery of frozen fish to be placed in the cold store.

The supervisor is the 'source' of information on the desired location for the fish. He translates (or 'encodes') this into words that are delivered as a 'message' through an appropriate channel. The channel in this case is a face to face conversation with the truck operator. The operator then has to convert (or 'decode') the message into a form which he (the 'recipient') understands. In this situation the instruction is on where to store the fish. This entire process can be described by a standardised flow diagram as shown in *Fig 2.1.*

Even in this simple example, there are numerous sources of error which can result in the truck operator not placing the fish in the desired position in the warehouse. One possibility might be due to 'selective inattention', *ie* the recipient does not actually receive the information correctly. For example, the operator's lunch break is almost due, in which case his thoughts are focussed on what to eat. Thus he does not accurately decode the instructions and unloads the fish at the wrong location.

An error may also occur if the supervisor is a 'distrusted source' because he is known to be indecisive, frequently changing his mind after the shipments have been placed at a specified location. A third common communications problem is the erroneous decoding of the message by the

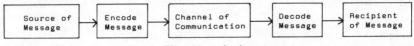

Fig 2.1 The communications process

recipient. This is most likely to occur if the source uses language which is not understood.

Once a group of employees have been working together for some time, they develop a common field of experience and everybody decodes messages correctly. A new individual joining the group has yet to acquire a full understanding of the format of standard messages. For example, as the warehouse supervisor might say, 'Put that fish with the other cod fillets'. This instruction will cause no problem to a new truck operator as long as there is only one type of code fillets in the warehouse. If the company buys both cod blocks and IQF cod, there is a 50% chance that the new shipment will end up being stored in the wrong place.

Effective communication

Effective communication requires a smooth flow of information from the source, via the selected channel (*eg* a letter or a telephone call), to the recipient. It is therefore necessary to carefully assess the following critical elements prior to commencing the communications process:

– Who is the recipient and how will they react to the information?
– Why are you communicating and what do you wish to accomplish?
– What precisely is the message that you wish to get across to the recipient?
– How do you wish to present this information: (*ie* if there is more than one point which you wish to communicate, how will you give priority to the points to be covered?
– When is the appropriate time to communicate?

Developing an effective communications style does not usually come naturally to most people. They have to work on their communications skills over time. It is very difficult to use dialogue as the starting point in the skills improvement process, because one has the complication of having to modify statements in response to feedback from the recipient. It is simpler, as the starting point for improving the information provision process, to use a written form of communication such as a letter.

Blue Fin – writing for results

The Blue Fin material provides a basis for examining actual practice and identifying approaches which can improve written communications. It is apparent that John Dennis is extremely frustrated, and possibly it would have been better to have handled the whole affair in a discussion, behind closed doors, with Alan Balfour. The latter was forced into responding in writing, because having received a complaint about his department, John Dennis was unavailable for any discussions until later in the week.

Starting with the Dennis letter, he did select the right person with whom to communicate. However the message he was trying to get across involved attempting to get a positive action from another person. This is often known as a 'DO' message. It should be borne in mind that effective 'DO' letters should:

- Attract attention to gain the interest of the reader and encourage the person to read on.
- Present a proposition in a way which indicates there are benefits for the reader, thereby persuading him to act accordingly.
- If possible, include an incentive to stimulate the recipient to react promptly.

The reason for the Dennis letter is to gain co-operation from Alan Balfour in revising the credit approval procedures in order to generate increased sales. What John Dennis needs to get across is that if the approval problem cannot be resolved, the company faces a further period of poor financial performance. By understanding these factors and by applying the guidelines for a 'DO' letter, it is possible to rewrite the letter as follows:

Interoffice memo

To: Mr A Balfour   From: Mr J Dennis
Accounting Director   Sales Director

Over recent months you have quite correctly in your financial performance reports been highlighting the need to increase sales[1]. As you are aware, a critical element of the company's current marketing plan is the expansion of sales in the retail sector using the new tray pack products.

The depot managers have encountered a very positive response from the retail sector, but there is a minor obstacle: namely obtaining credit approval for these new customers. This matter should really be resolved as soon as possible in order to fully exploit the potential incremental sales that can be generated.

Given the workload facing your staff, one possible approach might be to involve my depot managers in the information collection process[2]. At the moment your credit department staff contact (a) the new customer to obtain a bank and trade reference, and (b) one of the credit agencies for a credit search. May I suggest that specifically for the retail sector of the market, my depot managers ask for bank and trade references upon acceptance of the first order and also put into motion the check through the credit agency.

As you are aware, retailers tend to have a lower business failure rate and are prompter payers than certain other sectors of the food industry. Hence as retailers become a larger element of our customer mix, the successful expansion of our retail tray pack business should help to reduce the average age of outstanding receivables[3]. In view of this situation, I am also wondering if our depot managers can be given delegated authority to approve credit on the initial order up to a defined maximum level, from those retail outlets in their area which are a branch of any national chain. This would assist in achieving a more rapid market penetration for the tray pack products, thereby meeting the corporate objective of increased sales.

I am sorry it is not possible to be able to discuss this matter with you in person, but I have to be away from the office for the first part of the week.

Statement ([1]) is designed to gain Mr Balfour's attention by endorsing the correctness of his actions in identifying the need for improved sales performance. Statements ([2]) and ([3]) both offer benefits as the basis for gaining co-operation from the credit department in speeding up the credit approval process. The first benefit is the reduction in workload and the second, a reduction in the age of outstanding receivables. There is also an implicit incentive for rapid action in statement ([3]), namely that any procedural change will cause the receivable situation to improve along with corporate performance.

Alan Balfour's original letter was designed to say 'NO' and even this type of communication can be made more effective by covering the following issues:

- To ensure initial agreement from the recipient, open with a neutral statement.
- Present the bad news of 'NO' in a positive and tactful fashion, avoiding any phrases which imply criticism.
- To avoid any misunderstanding, state the bad news clearly and concisely.
- Suggest alternative actions, if possible, which may reduce the impact of the 'NO' message.
- Attempt to present all statements in a neutral or conciliatory fashion.

By applying these concepts to the Balfour letter, the response could have been presented as follows:

Interoffice memo

To: Mr J Dennis   From: Mr A Balfour
Sales Director   Accounting Director

Thank you for bringing to my attention the issue of credit approval for new customers for the tray pack product[1]. I am glad to hear that the product concept is gaining such an immediate acceptance in the market place.

As you are aware from the last Board meeting, it was decided that priority must be given to speeding up the collection of outstanding balances on customer accounts. In meeting this objective, I regret that the credit control department have less time available to handle credit approval requests[2]. At the moment, unless I can persuade the Board to approve additional staff, I cannot see a mechanism whereby there can be any change in the approval process.

Nevertheless I do realise the necessity for all departments to work together to ensure that the tray pack launch is a success. Perhaps, therefore, we could meet later this week to examine the issue. It might be possible, for example, to give priority to approval requests from the retail sector for the next few weeks[3]. This might cause additional delays on new customers in the catering sector. Hence it would be necessary to gain your perspective on whether this might be a problem for your sales staff.

Statement ([1]) is the neutral opening designed to gain initial agreement from Mr Dennis. Statement ([2]) says 'NO' but in a way which indicates Mr Balfour is acting to fulfil actions requested by Blue Fin's Board of Directors. To further reduce the impact of the 'NO' message, statement ([3]) is used to propose a possible alternative action which might reduce the credit approval time for customers for the new tray pack product.

Before sending any letter or memorandum, it is worth reviewing what has been written to ensure that there are no inaccuracies or typographical errors. There is a tendency in business to assume that complex phrases and 'jargon' can improve the credibility of written communication. It is worth remembering the acronym of KISS for all forms of communication: namely 'keep it simple stupid'.

One way of checking whether a letter fulfils the KISS rule is to apply Gunning's 'Fog Index'. Take the draft document and for about half a page of text, count the number of words of three or more syllables. The 'Fog' score is calculated by dividing the 3 + syllable word count by the number of sentences examined. If the score is found to be five or higher, it is likely that the letter contains excessively complex phrases that will not be understood by the reader. Under these circumstances, it will probably be worth editing the letter to (a) remove unnecessary words, (b) replace multi-syllable words with simpler ones, and (c) break long sentences into two or more shorter ones.

Communication through presentations

Sales staff are expected to be effective oral presenters because this medium is used to persuade customers of the benefits of an organisation's products or services. Developing presentation skills is a standard element of most sales training programmes.

Managers in other departments of an organisation are also frequently required to communicate to groups. Examples include a production director seeking approval from a Board for capital expenditure on new manufacturing facilities, an accountant reviewing a new control procedure to the company audit group, and a personnel manager briefing employees on a change in their company's pension scheme. Until recently, however, few organisations recognised the necessity of providing training on presentations for all managers, instead of restricting this type of instruction to the sales department.

Executives who have little trouble in using the telephone or a letter as a method of communicating, often approach a presentation with trepidation. They become concerned about issues such as not gaining the attention of the audience or failing to cover a key area persuasively. As with other forms of communication, however, by understanding and applying certain basic rules, most managers can soon overcome their initial fears and begin to use the medium of presentations as a very effective form of communication to gain approval of their proposed path of action.

Careful preparation is vital if the manager is to put across the desired message in a presentation. Although one is occasionally asked to make a presentation at short notice, in most situations the manager will be given prior warning. There is then time for a thorough preparation and rehearsal. This rehearsal should attempt to simulate the actual event, and it is therefore extremely beneficial to recruit an audience who would also be willing to offer constructive criticism to the presenter.

The objective of a presentation will usually determine its content. If the objective is to impart information, the content will focus predominantly on factual material. An objective of gaining approval or action from the audience will require ideas linked to reasoned argument supported by personal statements by the presenter. This latter element is the vital component in gaining acceptance by the audience of any key recommendations or requests.

Where the objective is to change behaviour in the audience (*eg* to persuade employees to accept a change in working conditions), the content of the presentation will have to accurately reflect the attitudes and beliefs of the audience. The manager will have to concentrate on ways of motivating the group to act, and must also be prepared to handle attempts by members of the audience to reject his proposition.

Careful thought must be given to the knowledge level of the audience. Attention can quickly be lost if the information presented is already well understood. At the other extreme, the presenter will lose his audience just as swiftly if they have only minimal knowledge and he moves straight into complex statements without first establishing some fundamental principles. Where an audience is known to be mixed, the manager will have to take a central position on the existing knowledge level within the group.

The size of the audience will significantly influence the level of formality of the presentation. With a group of ten or less, the presenter can encourage audience participation and thereby gain feedback on how effectively the information or required actions are being accepted. With a larger audience, the presenter will be forced to use a more formal structure, and use questions from the group at the end of the presentation to assess the audience response to his proposition.

Formality will also be influenced by the nature of the audience. When the presenter is addressing individuals from inside or outside the organisation who are senior in rank, it is preferable to create an atmosphere of deference and respect to them. A very common error of the inexperienced manager who knows some of the senior staff in the audience, is to use their first names when responding to a question. This is dangerous because it may be interpreted as showing insufficient respect and thereby reduce the presenter's credibility. When a manager is uncertain about the degree of formality expected, it is safer to err on the side of conservatism and be completely formal throughout the entire event.

The golden rule about the structure of a presentation is:

– Tell the audience what you intend to tell them (*ie* review the agenda).
– Make the presentation.
– Then summarise again for the audience what you have covered (*ie* re-review the agenda).

Highly complex data and detailed numerical information have a tendency to send audiences to sleep. Where such issues have to be covered, it is worth considering preparing a written report to be given to all the participants at the end. Never let the audience have this report at the beginning of the event because they will tend to spend their time reading the information rather than paying attention to the presenter.

There are a number of audio-visual aids available to assist the presenter (*eg* video, slides, audio tapes, flip charts). However these should only be used to enhance the information provision process, and not included simply because they are there. One problem with using video or film is that it can provide abrupt contrast between the skills of the live presenter and the professional narrator. Under these circumstances, the presenter may decide that the benefits of the film are greatly outweighed by the disadvantage of the audience making a comparison between the two levels of presentation skills.

The other drawback of certain audio-visual technology is that it has an embarrassing habit of breaking down. The presenter should ensure either that an equipment technician is on hand, or take along appropriate spare equipment to cover all contingencies.

In judging whether the audio-visual aids can assist the presentation, the manager should seek mechanisms which complement, not duplicate, his presentation. There is little point, for example, in using a set of slides to communicate certain facts and then making verbal statements which exactly duplicate these points.

Ultimately the quality of a presentation will be influenced by the confidence with which the manager presents the material. This is why it is so important to rehearse the event until the manager has an in-depth understanding of the information to be imparted. If possible, the manager should attempt to reach the point where he can communicate without any reference to a script. There is nothing more damaging to credibility than the reading of a presentation word for word from detailed notes. The audience reaction is that they could have been sent a report instead of attending the event.

If the manager cannot become word perfect, then the usual solution is to use small index cards on which there are prompts about the issues to be covered. The cards can be held discreetly in the hand or placed on the podium in front of the presenter. It is important to stress that the presenter should only glance down at the next card for the shortest possible moment, and concentrate at maximising the time when he is looking out at his audience and sustaining eye contact with them.

One of the most effective ways of improving presentation skills is to see yourself on video, recorded either at rehearsal or at the actual event. For most of us, there is nothing more agonising than seeing ourselves on video, but as few of us are natural presenters, this is an incomparable tool for developing a more professional presentation technique. Every manager should endeavour to see themselves through the medium of video at the earliest possible point in their career.

3 Motivation

Case: Marine Farms Ltd

David Gallon joined Marine Farms as a technician working on the research and development programme to establish a shrimp farm. His main area of responsibility was to contribute to the establishment of the hatchery. Within two years of his joining the organisation, the R&D had progressed from the experimental phase through to scaling up to commercial production. At this point, the research team was disbanded and David was appointed as Hatchery Manager. This change was accompanied by a 25% increase in salary.

David was made responsible for ten subordinates and reported to the company's Production Director. Bringing the full scale hatchery on-stream was very stimulating, but as the operation became a matter of routine David began to find the excitement had gone out of his job. Although the hatchery was efficiently operated, his colleagues and subordinates noticed that David's enthusiasm for his work seemed to have declined.

Performance at work

For many years behavioural scientists have studied the question of why people behave as they do at work. Their objective has been to understand what causes people to perform effectively. These studies have centred on two theoretical concepts – motivation and behaviourism.

The two main concepts are based on different explanations of how people perform. Motivation theory makes assumptions about the mental state of employees and the influence on performance of the individual's desires or wants. Behaviourism ignores inner needs and explains employee behaviour solely in terms of its consequences in the work environment.

The human animal is an extremely complex organism, so no single theory about motivation or behaviour can be used to solve every employee performance problem. With this in mind, the purpose of this chapter is simply to illustrate the application of various concepts for improving people's performance in an organisational environment.

In real life, any application of the various theories must be treated with extreme care. In complex situations, the manager is strongly advised to seek professional assistance from the company's personnel department or an external management development expert.

The hierarchy of needs theory

Motivation is a measure of the extent to which people commit themselves to achieving goals which satisfy their own needs. These needs are complex and can be satisfied both within and outside the work environment.

Maslow has classified people's needs into a pyramid or hierarchy. He suggests that as a particular need is satisfied, the influence of the next level of need in the hierarchy then emerges. In consequence, needs are never static; they change over time and are influenced by both experience and expectations.

As illustrated in *Fig 3.1*, the more basic physiological needs (*ie* food, drink, *etc*) are fundamental to everyday existence. Only after these have been met, can more complex motivations involving an individual's personal aspirations (*eg* esteem or self actualisation) begin to influence behaviour. The application of Maslow's theory is in understanding how the manager or his subordinates are motivated and how performance can be influenced. Work is not a motivator: it is what people expect to get out of their work which causes them to behave in various ways. An effective manager is one who creates conditions for himself or his subordinates which enable certain personal goals to be reached by linking these to the objectives of the organisation.

Applicable personal goals which can be incorporated into performance of benefit to the organisation are:

− a need for a sense of achievement;
− recognition of good performance;
− advancement and promotion;
− participation in decision making;
− increased responsibility;
− need to organise and plan one's own work activity;
− challenge and personal growth.

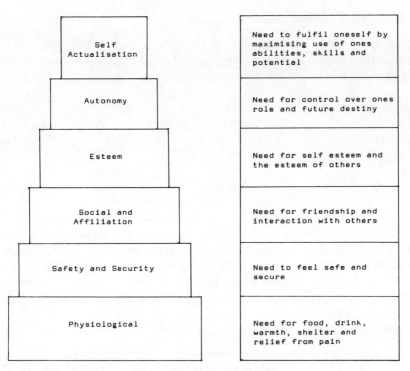

Fig 3.1 The hierarchy of needs

One application of Maslow's theory has been provided by Professor Porter who used a questionnaire to measure executives' response to conditions in, and their perceptions of, their work environment. The respondents were asked to classify a number of statements in relation to a scale which ranged from 'none at all' to 'a great deal'. For example:

– 'How much self esteem *should there be* for a person in my job position?'
– 'How much self esteem *is there* for a person in my job position?'

The research revealed that the pattern of needs in the average organisation was, in descending order: self actualisation, autonomy, social, security and esteem. This means that the higher level needs in the Maslow hierarchy are more important to the average executive. Furthermore, as executives move into higher levels of management, a very pronounced importance develops for self actualisation and autonomy.

Marine Farms – role perception

Application of a method such as Porter's to ascertain David Gallon's perceptions of his role during the research/start-up phase as compared to his present role might generate the following scores:

Need type	Research phase	Current role	Change
Security	4.80	5.60	+ 0.8
Social	5.60	4.90	– 1.3
Esteem	5.30	5.50	+ 0.2
Autonomy	5.60	5.00	+ 0.6
Self actualisation	6.40	5.20	– 1.4

The situation facing David is not uncommon – that of having moved from a flexible, unstructured, R&D environment to a role of managing staff in a structured, production-orientated, rigid cost and time controlled environment. During the research phase, he was meeting his self actualisation need by contributing to the solution of complex scientific problems. This sort of opportunity disappeared as the hatchery operation became more established.

Possible solutions which David and his boss could examine are:

- Methods for re-introducing some aspect of research into the job role (*eg* giving authority to examine alternative culture techniques for the plankton used as the feed during the early post-larval phase).
- Providing David with further training to help him recognise the importance of his new role within the organisation and develop new managerial skills (*eg* leadership, communication, planning and control). This should be accompanied by counselling on the promotional and career opportunity implications available as the company continues to expand.
- Agreeing that his future might be better served by moving to a more research orientated environment. This might be within the company, but if this is not feasible to actually assist him in looking for a new job.

The hygiene/motivating factors theory

Extensive research by Herzberg has led to the suggestion that two types of factors have to be satisfied if people are to be motivated to perform effectively. The hygiene factors (or 'dissatisfactors') do not cause employees to do better, but if they are not accommodated, individuals will not work very hard. The way to influence performance is to provide additional elements – the motivating factors (or 'satisfiers').

Herzberg considers issues such as company policy and working conditions as dissatisfiers because they are found to be the cause of discontent if not correctly handled. He also includes administration methods, supervision, relationships with superiors, associates or subordinates, security, status and salary as sources of potential dissatisfaction.

Having minimised the influence of dissatisfiers, managers then have to ensure the presence of satisfiers in order to enhance performance. Issues such as designing the job to yield feelings of adequate responsibility, achievement, personal growth and recognition are among the more usual forms of satisfiers.

Another perspective on the Herzberg approach is to consider the dissatisfiers as those needs which comprise the lower levels of the Maslow hierarchy. These must be fulfilled before one can use the higher level satisfiers of self actualisation or esteem to enhance performance.

Measurement of hygiene and motivating factors is achieved by asking employees to identify the degree of importance they attach to various issues in the work environment. For example, questions on hygiene factors may ask respondents to rate the degree of importance they attach to:

- Constructive supervision
- Pension scheme
- Job security.

These questions would be mixed in with statements concerning motivating factors such as:

- Recognition of good performance
- Opportunities for advancement and promotion
- Freedom to organise one's own work task.

The Herzberg theory has been widely used to examine employee performance and the possible need to restructure specifications of job roles within organisations. However, it is as well to be aware that the theory has been considered controversial. For example Herzberg classified salary as a hygiene factor, *ie* financial reward would not motivate employees to improve their performance. Some other management development experts consider salary as a motivating factor; and others classify it as both a potential satisfier and dissatisfier. There is also evidence that factors will vary in importance in different work situations (*eg* employees packing fish fillets versus clerks in the accounting department), and between different people doing the same type of job. Nevertheless, as a concept, if the analysis is handled correctly, the Herzberg theory has been proved to be a very practical tool for managing motivation.

Marine Farms – motivation factors

Applying the hygiene/motivating factor concept to David Gallon, for example, we might find that out of a maximum of 100 the two scores for his current job would be:

Hygiene factors	*Motivating factors*
70	40

It can be seen that the hygiene factor score is relatively high compared to that given for the motivating factor. This would lead to the conclusion that he is motivated to perform well under current work conditions but, as with the earlier hierarchy of needs analysis, changes will be necessary if David is to be motivated to improve his overall performance.

Expectancy theory

The Maslow and Herzberg approaches focus on the issue of 'what' motivates people. Researchers such as Vroom have examined the process of motivation by studying 'how' people are motivated. This approach has led to the development of expectancy theory models which usually contain the following three elements:

– *Expectancy* – the belief that an action (*eg* additional effort) will result in a known outcome (*eg* improved performance).
– *Instrumentality* – the issue of whether the improved performance will provide a benefit (*eg* higher earnings or enhanced status in the organisation).
– *Valence* – the issue of whether the identified benefit is of any value to the individual concerned.

In applying this theory on motivation, the manager must determine whether it is possible for the employee to perceive a direct relationship between an action and a specified outcome. For example, a processor who wants to increase the average fillet yield per pound of round fish would have to persuade his filleters that if they were more concerned about yield and concentrated less on maximising the pounds of fish filleted per day, then yields could be increased. This may involve some retraining as well as presenting information from the daily production records to prove to employees that by taking slightly more time per fish, yields can be increased.

Being able to prove that yields can be improved through a minor change in the work process will have little influence on performance unless the elements of instrumentality and valence are also fulfilled. The benefit to the company of higher yields would be to increase overall profitability. This is not likely to be seen as very important by the filleters. Hence the management will probably have to examine a mechanism such as a weekly productivity bonus based upon higher yields. This bonus would then fulfil the instrumentality element because employees will perceive a direct benefit to themselves in meeting the request to increase yields. However for this bonus to motivate the employees, it must be perceived as having a significant value (*ie* the valence element). This will probably mean that the productivity bonus would have to be able to increase earnings by at least 2–4% above current pay levels.

Behaviourism

Theories of behaviourism ignore concepts about needs or inner mental states, and merely focus on the concept that certain behaviour will elicit a certain reaction. A simple example is that when a baby cries, the consequence is usually to attract the parent's attention. The arrival of the parent in reaction to the cry is recognised by the child and strengthens this crying behaviour pattern for gaining attention in the future. This process is known as operant or instrumental conditioning.

It is thought that there are four basic consequences which can influence behaviour through operant conditioning. In organisations, their form (and examples of each) are as follows:

- *Positive reinforcement* – such as paying a bonus for higher output.
- *Negative reinforcement* – where an employee who is anxious about an interview finds the anxiety reduced by involvement in the interview process.
- *Omission* – where an employee's willingness to work harder is reduced upon realising there are no bonus payments for extra effort.

– *Punishment* – as exemplified by making a deduction from an employee's salary for frequent late arrival at work.

Applying behaviourism in the work environment should be handled with extreme care because the operant conditioning theories are based upon laboratory studies of animals. Thus translating these theories into man-management systems outside a strictly controlled laboratory situation is not only difficult, but can be extremely risky if attempted by managers with little formal training in human psychology. Nevertheless when utilised correctly, the concepts have produced some well documented examples of success (*eg* application within the Emery Air Freight Corporation in America).

Employees exhibit two basic types of behaviour – desirable and undesirable. The objective of the manager is to reinforce the desirable and eliminate the undesirable. This can be achieved through controls based upon aversive (*ie* punishment or negative reinforcement) or affirmative actions (*ie* positive reinforcement or omission).

The natural inclination of most managers is towards the aversive control of punishing bad performance. Management research studies, however, have shown that this type of control is only effective over the short term. Successful examples of the application of behaviourism have typically centred upon affirmative forms of control.

In designing a suitable form of positive reinforcement, consideration has to be given to its scheduling. A continuous schedule requires that the reinforcement is provided every time the behaviour occurs (*eg* an electronics technician is paid a bonus each time he completes a service on a Decca Navigator on the company's trawlers in less than the allocated time for such work). Although continuous reinforcement will help speed up learning, the provision of immediate 'on the job' reinforcement is usually very difficult to achieve in most organisations. There is also evidence to suggest that if there is any delay in this type of reinforcement (*eg* the wages clerk is delayed in issuing the bonus payment), then employee performance will rapidly decline. Consequently most organisations use intermittent reinforcement schedules.

An example of an intermittent schedule would be the payment of a productivity bonus at the end of each week of production. This approach is known as a fixed schedule because the employees are aware of when they can expect to be paid. This can be contrasted with a variable interval form of reinforcement such as a supervisor dispensing a compliment for good performance, which tends to occur on an irregular basis which cannot be predicted by his subordinates.

Application of behaviourism

The majority of applications of operant conditioning occur as an unplanned aspect of managing employees because the good manager naturally utilises techniques such as praise to further enhance performance. The important aspect of unplanned conditioning is to recognise that positive reinforcement is to be preferred over the negative. Although an action such as punishment of an employee seems to eliminate problems, it should be realised that over the longer term this will probably generate adverse side effects. Employees told off for excessive talking during working hours, for example, just learn to contain this behaviour when the manager is in the vicinity. As soon as the manager has left, conversation recommences with a consequent reduction in productivity.

Where a company decides to introduce a structured programme of behaviour modification, it will usually involve completion of the following steps:

- Identification of the area where current behaviour is reducing productivity.
- Measurement of the current level of productivity and determination of the desired level of improvement. (This will frequently involve employee participation).
- Design the form of positive reinforcement and commence application.
- Re-examine productivity to determine if the reinforcement is achieving the desired level of improvement.

Case: behaviourism at Shrimp Farms Ltd

The majority of the output from Shrimp Farms' pond system is processed by approximately 50 women who hand-peel the shrimp tails prior to freezing. The Production Director felt that compared to output levels achieved in other plants in the shrimp industry, the peeling operation was not very efficient. His viewpoint was based upon both the weight of tails peeled per shift and the ratio of peeled meat versus weight of shell-on tails processed.

He therefore decided to attempt to improve peeling rates by 10% and to improve meat yields by at least 5%. To establish a form of positive reinforcement, the peelers were assigned a specific work table and placed into one of five peeling teams. In the past, the employees worked at whichever peeling table they pleased. Each team was identified by a colour (red, blue, green, yellow and orange) and each team elected their own leader. At the end of the peeling room, a large blackboard was installed. At the end of the morning and afternoon work periods, the output and meat yields for each team were written up on the board by their team leader. Natural positive reinforcement was provided by the employees seeing the performance for themselves. Contrived reinforcement was created by the production supervisors praising good performance and helping less experienced employees by showing them how to improve their peeling technique. Feedback is provided by both the tabulated results on the blackboard and the frequency of praise from the supervisors. After a few weeks the natural reinforcement of the displayed results was replaced by the reinforcement created by the teams competing against each other to achieve the highest output and yield results.

After only six months, the Production Director was pleased to find that output had risen by almost 13% and yield by 6%. Even more importantly, this enhanced level of performance showed no sign of any downturn.

An interesting side effect of the new system was that morale of other workers in the plant also improved as they became very involved in observing the output rates by the peeling teams. The outcome was that the other employees began to develop ways in which to help the peelers even more. For example the warehouse staff began to arrive 15 minutes early for work so that they could set up the boxes of shell-on shrimp at the end of each table. This meant that the peelers did not face any delay at the beginning of the shift through having to wait for raw material to be delivered to their tables.

4 Managing subordinates

Case: Clipper Products

Clipper Products is a port wholesaler which deals in fresh and frozen shellfish. Their traditional customers have been inland wholesalers, but recently Clipper has agreed to act as a direct supplier to a local supermarket chain.

Mr George Sharp, the Sales Manager, has realised that there is a need to create a new position in the organisation to liaise with individual store managers of the retail chain and the various departments within the company. He has decided to appoint Steven Williams to the position of Retail Sales Co-ordinator. Steven has worked as a senior accounting clerk with Clipper for approximately three years, so he already has a reasonable background knowledge of the company's products and operating procedures.

Steven has decided to accept this offer of promotion. The next task for George Sharp is to instruct his new subordinate on the various aspects of his job.

Briefing subordinates

Personnel managers and management consultants frequently find that an underlying cause of poor performance within organisations is that no one has (a) fully explained the specific job role to employees at the outset and/or (b) provided adequate training if the new employee lacks certain job skills.

This situation can easily be avoided if the manager carefully plans and implements a job briefing programme for any newly appointed or promoted employee. The first step in the briefing process is for the manager to break the job down into its component stages. Each stage should be closely analysed to determine how much detailed instruction will be required to explain the various elements to the employee. Consideration should be given to whether these instructions can be communicated verbally or whether there is a need for a written policy manual. In large organisations, there is now a trend towards the introduction of video-based instruction material and some companies are experimenting with the benefits of interactive learning systems as a key component of the employee induction process. Whichever method is to be adopted, the manager should endeavour to place himself in the position of the subordinate who will be receiving this information for the first time. This approach by the manager will then ensure that no information is omitted from the briefing system which might be vital to the subordinate's full understanding.

If there is any doubt concerning the ability of subordinates to understand any part of the briefing, the safest starting point is to assume that the subordinate has no knowledge at all. In this way, the manager will tend to cover all issues in depth and thereby avoid confusing subordinates because of an omission of information which has left them not really understanding the nature of their new job role.

Having decided which format to use for the briefing, it is a good idea to put the material to one side for a short time and review it again later. Any omissions or errors in the proposed briefing will be more readily identified at this time. It may also be useful to discuss the proposed instructions with a colleague to see if there are any areas were the instructions are incomplete. Alternatively the information can be reviewed with an employee who has been executing the actual job in question for some time.

Having finalised the agenda for the job briefing, the manager should develop a detailed list of the skills the subordinate needs in order to execute the job role effectively. This should then be compared with the employee's existing skills in order to identify any area where further training might be necessary. Finally the manager should determine what resources will be required (*eg* access to a telephone; secretarial support) for the subordinate and ensure that these are made available at the time he starts work in the new job role.

Delivering the employee briefing

It is vital that the subordinate is put at ease so that he is in a receptive frame of mind. The manager should select a location which ensures that they are not disturbed. The meeting should be scheduled to allow sufficient time for a detailed discussion of any issues the subordinate might raise during the briefing process.

At the beginning of the meeting, the manager should establish the reasons why this job is necessary and its relationship to other activities within the organisation. The component stages of the job should then be described in detail and the subordinate encouraged to stop the presentation at any point where he would like further clarification.

The manager should be prepared to go over the instructions a number of times if the subordinate shows any signs of not understanding the information. At the end of the presentation, the manager should probe the subordinate's responses to check that the briefing has been fully understood. Finally the manager should describe clearly the performance that will be required of the subordinate and explain why the company has selected him for this particular job.

The subordinate should then be asked for their opinion on whether they feel they can handle the job or if there are areas about which they are concerned. The manager should treat the latter type of response as a positive contribution from the subordinate and be prepared to discuss how these concerns might be eliminated (*eg* sending the employee on a training programme; involving a more experienced subordinate to provide temporary support during the learning phase).

The manager may find it beneficial at this point in the meeting to have the subordinate run through the job tasks to check that they have understood the briefing. Subordinates will sometimes not be willing to admit ignorance or any lack of understanding in case this is interpreted as an indication of incompetence. This reticence will usually be avoided if the manager has (a) made an effective presentation on the tasks involved, (b) communicates a patient, constructive attitude when dealing with questions, and (c) makes it clear that the subordinate is not expected to be 100% perfect in the execution of the new job immediately he starts works.

Once the subordinate has begun executing his new tasks, the manager should carefully monitor performance, provide encouragement and, when appropriate, constructive guidance on how to overcome any problems. Supervision of the subordinate should be done in an atmosphere which encourages the employee to ask for help when necessary and creates a willingness to propose alternative procedures which might improve productivity.

Clipper Products – briefing subordinates

Prior to briefing Steven, George Sharp developed a flow diagram detailing the specific aspects of the sales co-ordinator's role (*Fig 4.1*). To assist the briefing he prepared a folder for Steven which contained the flow diagram, a sample copy of the company's inventory report, a shipping instruction

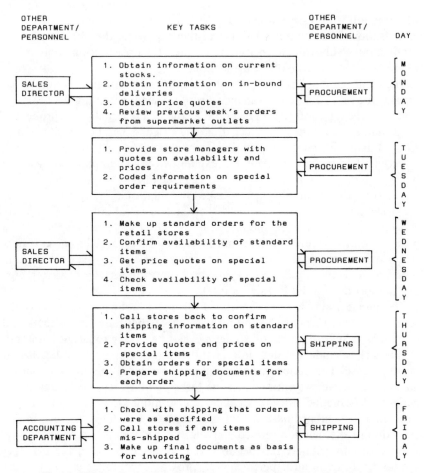

Fig 4.1 Clipper Seafoods – flowchart of job tasks for subordinate briefing

form, example invoice and a complete list of the store addresses, managers' names and telephone numbers.

The agenda for the briefing was divided into six stages:

– Review of the job role.
– Question and answer session on the job role explanation.
– Brief presentation by the Procurement Manager.*
– Brief presentation by the Shipping Department Manager.*

- Brief presentation by the Head of Accounting on invoicing procedures and credit control.*
- Steven's review of the procedures followed by further questions and answers between him and George.

(* These three individuals would only be present at the meeting during their briefing phase.)

George scheduled the briefing for a Friday afternoon because this was a time during the week when he knew he was unlikely to be disturbed. The meeting with Steven started at 2 p.m. and lasted about two hours. The briefing went well, so George made it clear he would leave Steven to start work in the new role with the minimum level of supervision. He observed that Steven was exhibiting enthusiasm and effectiveness, points which he praised when they met at the end of the first week to review progress. At that time, Steven had a couple of queries about procedures and George spent approximately 20 minutes clarifying these for him.

Selling direct to the supermarket chain proved to be very successful and George Sharp was pleased to be approached by an even larger retail group to also supply their stores. He discussed the situation with Steven Williams who felt there would be no problems servicing this new customer. Mr Sharp monitored the situation for a few weeks to ascertain whether the additional workload was impairing Steven's effectiveness. It became apparent that the preparation of invoices and shipping documents was putting pressure on Steven, so he arranged for one of the accounting clerks to provide assistance on a part-time basis, for two days during the middle of each week.

Steven had always had a reputation for being good humoured and co-operative. George Sharp was surprised to notice some three months later, therefore, that Steven had become less cheerful. At first George thought it was a passing phase, but after a week during which Steven clearly wanted to keep all conversation to a bare minimum, he realised that there was a problem which needed handling.

Recognising that a subordinate has a problem

A manager must be continually on the lookout for any indications that an employee has a problem. The problem may be with their work or in their personal life. Clues provided by employee behaviour include being uncommunicative, sulking, arguing with colleagues, superiors or subordinates, showing irritation, absenteeism, late arrival at work, and working at a much slower pace than usual.

A subordinate with a problem will usually be less productive and it is therefore vital that the manager act immediately to resolve the problem. In addition, the poor performance of one subordinate will influence the performance of other employees who rely upon the individual for information, action or response. There is a further risk that the subordinate may eventually become so frustrated that they actually damage company operations (*eg* by being rude to a customer; being careless and causing an industrial accident). This type of outcome is unintentioned by the employee, although employees with a problem can sometimes wilfully attempt to damage the company (*eg* sabotage or fraud).

Reacting to the problem subordinate

Problems with subordinates can arise because of unsatisfactory conduct and/or inadequate performance. Even if an organisation has formal disciplinary procedures, the manager should be prepared to become involved and not immediately refer the whole matter to the company personnel department.

It is of great benefit to a manager if he sets a good example in relation to observing all company policies and procedures. In this way, he will avoid subordinates breaking rules based on the justification that the manager behaves in the same way. It is also important that when a problem becomes apparent, the manager should act immediately. Any delay in dealing with the matter will usually cause the problem to become larger. The delay may also be seen by other subordinates as either weakness or favouritism.

When a manager is confronted by a problem subordinate, it is preferable not to handle the situation in front of other employees. The subordinate should be taken to one side and the issue discussed in private. Throughout the whole process of attempting to resolve the problem, the manager must retain self control and act calmly no matter how irritating he finds the subordinate's attitude or behaviour.

Dealing with the subordinate's problem

In many instances, the subordinate may know he has a problem but does not understand or recognise the underlying cause. Furthermore, even when the manager believes he knows the best solution, he should avoid trying to impose this on the employee. A more constructive approach is to work with the subordinate to help him come to understand his own problem and where possible, define an appropriate solution for himself.

When discussing a problem with the subordinate, the manager should be aware that employees often do not feel free to express their real feelings on the matter. This is due to the fact that very open statements might annoy their superior. Hence the manager should endeavour to create an interview environment which places the subordinate at ease and permits the manager to give his whole attention to the discussions without any outside interruptions. Frequently the subordinate is expecting to be greeted with anger, and the manager's opening comments should be designed to communicate that he wishes to constructively assist the employee to resolve the problem. Caring, positive phrases should be used, such as 'Can I be of help?', 'Would you like to talk about what's bothering you?' or 'Talking things over might really be of benefit'. At this point the employee needs a listener, not a lecture, and the manager should give the individual a chance to explain fully what is bothering them. This phase of the discussion can be assisted by the manager confirming that his intention is to gather information, and that he is not simply waiting for an opportunity to jump in with his own opinion. This can be achieved by the manager nodding, maintaining eye contact and using encouraging phrases such as 'Go on' or 'I understand'.

Gathering information about the problem

It is very likely that the subordinate will not provide sufficient information to permit the manager to contribute to the development of a solution. Under these circumstances, the manager should be prepared to ask questions.

Effective questioning is not as simple as it may seem because there are a number of potential pitfalls. The questioner must avoid phrasing the questions in such a way as to risk generating a biased response. All that happens then is that the questioner receives invalid information. One form of biasing is when the manager uses the question format of 'You do agree of course that........is true?'. The manager is indicating that he already holds a strong opinion on the issue, and the subordinate will usually agree in order to avoid a confrontation with his superior. Another potential pitfall is where the manager phrases the question so that it is not understood. This can occur if the manager uses words that are unfamiliar to the subordinate or by making a very lengthy statement which contains a whole series of different questions.

The most effective way of generating information is to ask questions in the most appropriate form. There are three basic forms of question: open-ended, closed and multiple choice.

Open-ended questions are appropriate if the manager wants to generate a broad ranging response from the subordinate (*eg* 'What is your opinion about......?'). This type of question is also very useful at the beginning of a discussion when the manager is trying to develop a rapport with the respondent. Closed questions ensure a very specific answer. They are applicable when the manager wishes to confirm a precise point (*eg* 'Is the increase in costs greater than 10%') or collect a specific piece of information (*eg* 'How much will it cost per year to lease that item of equipment?'). Where there are a range of possible responses, but the manager wishes to partially restrict the breadth of response, then a multiple choice format can be used. For example, 'Which are the three most important issues to be considered in selecting the new machine: cost of operation, servicing needs, available space in the plant, amount of training required by new operatives, compatibility with current equipment, or availability of spare parts?'.

It is sometimes beneficial to prepare a list of information that will be sought from the subordinate and the questions to be asked. Then if the questions in whatever format do not provide all of the answers, the manager can pose some follow-up questions. These will have the benefit of stimulating the subordinate to continue offering his opinions (*eg* 'How do you think we could go about obtaining the additional information on operating costs?'). In consequence, the manager will be able to gain a deeper insight into the subordinate's understanding of the various issues under review. Follow-up questions are also useful in confirming the manager's understanding about the information already provided (*eg* 'Does that mean our costs are going to increase by over 10%?') or gain confirmation on the agreed next actions (*eg* 'Do you feel we can implement the new cost control system before the end of the year?').

During the questioning phase, the manager should endeavour to concentrate on the subordinate's responses by listening carefully to what is being said. When the manager thinks he knows the answer or he finds the pace of the conversation too slow, there is a tendency to think more about his next question than the present response. This may cause the manager's concentration to waver and miss an important fact. More importantly, if the subordinate realises he is not getting 100% of the manager's attention, this will defeat the objective of involving him in working with the manager to develop an appropriate solution.

Where the subordinate is stating facts or expressing an opinion over issues that are obviously very important to him, the manager should underline his receptiveness by using responses which paraphrase the subordinate's statements. For example the subordinate may be having a problem justifying that there is a variance in actual performance versus budget.

Subordinate: 'I do not feel I am getting the co-operation from the accountancy group on accurate cost figures'.

Manager: 'So you believe that the cost figures are vital to the situation and you feel the accounting department is not co-operating over this issue?'.

The solution framework

The process of getting subordinates to solve their own problems with a minimal level of input from the manager can be greatly assisted if the manager directs the discussions along a structured path. A logical and usually effective approach is to break the process into the following sequential steps:

- Identify and define the problem.
- Generate sufficient information to provide the basis for a number of alternative solutions.
- Analyse the strengths and weakness of the various alternatives.
- Select the most appropriate solution.
- Plan the most effective implementation of the solution.
- Implement solution and determine a point in the future when the viability of the solution can be assessed.

Clipper Products – handling a subordinate's problem

George Sharp decided that it would be better to hold the meeting with Steven away from the office and under conditions where Steven would be relaxed. So he invited Steven out for lunch at a quiet restaurant where he knew they would not run into either customers or other Clipper Products employees.

George explained to Steven that he was concerned about his performance and asked him to try to explain in detail what was bothering him. It emerged that Steven was finding that the expansion of the sales to the new supermarket chain seemed to place him in the situation where everybody was blaming him for problems that occurred. The store managers were complaining that Clipper Foods never had enough stocks and were slow in responding on prices for special items. The delay in confirming prices meant the Shipping Department were not receiving the final shipping information by Thursday. The Accounting Department was complaining that Steven was making too many demands on the accounting clerk allocated to him to assist with the paperwork. George made it clear that Steven was thought to be doing an excellent job and that these complaints were probably due to situations currently outside Steven's control. He therefore suggested that Steven might like to propose some solutions to the various problems which seemed to be bothering him. Much to George's relief, Steven was very willing to analyse some possible solutions and they continued the discussion for the rest of the afternoon back at the office.

The outcome of these discussions was that:

- Steven would carry out more detailed analysis of the stores' needs through meetings with their managers, and use the data on earlier weeks' sales to examine sales trends.
- Procurement would be instructed to keep Steven updated on catch and landing information so he would have a better understanding of availability when talking to the store managers.
- Steven would prepare monthly estimates of the store needs for review with Procurement. They would be able to use this information when placing orders and if there was going to be any shortfalls, could alert Steven so that he could pass this information onto the store managers.
- Getting prices on special orders was delayed because George was not always around when Steven needed an answer. George therefore would instigate a weekly meeting with Steven to discuss prices, and Steven would be given authority to set prices for the stores without checking with George.
- The increased workload facing Steven meant that he needed a full time assistant who reported to him rather than to another department head. George undertook to either arrange a transfer of the accounting clerk to work under Steven on a full time basis, or recruit a new member of staff from outside the company.

5 Managing through others

Case: Avon Foods Ltd

Avon Foods manufactures frozen, breaded fish and seafood products. The Quality Control Department at the processing plant is staffed with four production supervisors, two laboratory technicians and a food technologist. Their activities are directed by Mr Ralph Peters, the Quality Control Manager.

One Monday morning, the Sales Director Charles David attempted to inform Mr Peters that a major chain of fast food restaurants wanted Avon Foods to bring forward the plans to produce a 3 oz breaded fish wedge. Prior to placing their first order, the customer wanted to assess output from a pilot production run. Samples from this run would be sent for trials at a restaurant. With the samples, they also required a complete quality control report describing average breading levels, uniformity of shape, breading colouration and moisture levels of the fish.

Mr David discussed the matter with the plant manager who told him that although he could accommodate the request for a trial production run, there might be a delay in (a) drawing up the final production specification and (b) completing the quality control analysis. The reason for the delay was that Ralph Peters was away all week, attending an important Government conference on possible changes in the legislation for the labelling of fish and seafoods. Mr David, therefore, asked Mr Peters' secretary to contact him at the conference and explain the urgency of the matter.

The next morning Mr David received a telephone call from the secretary to tell him that Mr Peters would not be back until the end of the week. Hence she was very sorry, but the samples could not be produced until her boss returned.

Approximately an hour later, Mr Peters was called out of his conference for an urgent telephone call from Avon's Managing Director, Ronald Swan. The following conversation ensued:

'Morning Ralph. Charles has just explained to me that there might be a slight delay in delivering the pilot production run samples of the new 3 oz wedge. As you know, I only get involved in situations if I think I can help. In this case, my advice to you is quite simple. Either figure out how to be in two places at once, or that conference will have to proceed without you.'

Being in two places at once

All managers face excessive demands on their time in getting things achieved. As there is never enough time in the day, the effective manager can only fulfil all his obligations by involving his subordinates through the process known as 'delegation'.

As discussed in chapter 1, however, many individuals exhibit a major aversion to delegating important tasks. The reason for this is usually due to (a) an unwillingness by the manager to accept that subordinates can effectively carry out such tasks and/or (b), a lack of understanding of the actual processes associated with delegation.

To establish whether or not a task can be delegated, it is necessary for the manager to honestly review the following issues:

- *Saving time* – can the task, which the manager cannot handle until later be carried out immediately by somebody else?
- *Saving money* – is there a subordinate who is paid less than the manager, who can become involved and thereby reduce the cost of executing the task?
- *Superior expertise* – are there subordinates who have a specialist capability which means they can actually execute the task more effectively than the manager?

OR

- *Acceptable expertise* – are there subordinates who would not do the job as well as the manager, but whose performance will be adequate for the task in question?
- *Employee development* – would any subordinate benefit from involvement in the task, thereby gaining experience which would enhance their performance in the future?

It is usually found that most individuals who are averse to the thought of delegation, will be forced to respond 'YES' to a number of the above questions. Where this is the case, both the manager and his subordinates will obviously benefit from the manager expanding his subordinates' involvement in more tasks. The other beneficiary will be the organisation, because the manager will be able to become involved in other important matters (*eg* allocating more time to long term planning), and this will ultimately lead to an improvement in organisational productivity.

How to delegate

Effective delegation of new tasks requires careful preparation by the manager. Merely handing the subordinate a file with the instruction 'Need this done by Friday, call me if you have any problems', is guaranteed to ensure failure by the subordinate.

A more appropriate handling of delegation is one which involves:

- *Proactive thinking* – where the manager thinks ahead and does not wait until a new problem arises. Careful review of future events will lead to identification of issues which can be managed through a timely assigning of the task to a subordinate.
- *Task totality* – whereby the manager endeavours to allocate the entire task to the subordinate, rather than just one or two aspects of the job.
- *Appropriate nomination* – achieved by a careful review of the capabilities of available subordinates and selecting an individual on the basis of appropriate attitude, personality, existing and potential technical skills.
- *Task importance* – a process of determining how important is the delegated task relative to meeting key organisational objectives. Where a certain degree of failure can be tolerated without any damage to the organisation, then the manager can give consideration to using a subordinate who would benefit from involvement in activities outside their area of previous experience.
- *Objective setting* – through which the subordinate has a clear understanding of the objectives which need to be met in execution of the delegated task. The manager should be prepared to involve the subordinate in a discussion of the objectives and accept that these might be modified in light of the subordinate's perceptions of how best to approach the assigned task.
- *Flexibility* – a clearly communicated willingness by the manager to be prepared to accept modification of both time scales and objectives if the subordinate encounters unforeseen problems. This is especially impor-

tant when working with relatively inexperienced subordinates.

- *Equality* – the manager should not just delegate the boring or unexciting jobs. Subordinates will rapidly be demotivated if the manager is seen to keep all the stimulating and exciting tasks to himself.

- *Achievement recognition* – which requires that the manager communicates to superiors which subordinate was responsible for well executed assignments. Equally, the manager must remember that he is responsible for any failures. Hence when things do go wrong, the manager must accept the blame and not immediately claim it was the fault of somebody else in his group.

- *Patience and trust* – once a task has been delegated and a timetable mutually agreed, the subordinate should be left to get on with the task without the manager continually checking up on his progress. Even when the manager can see things are going wrong, unless it is imperative that the mistake is immediately remedied, the manager should patiently wait until the subordinate comes to him for guidance or help.

Avon Foods 'what should have happened'

The new 3 oz breaded fish wedge could have provided Ralph Peters with an excellent opportunity for delegation prior to departure to his conference. The food technologist Barry Evans works under Ralph's direction on new product development and reformulation projects on existing products. He therefore qualifies on the factors of delegation of (a) having the necessary expertise, (b) assisting his personal development as an employee, and (c) the organisation would save both time and money if he managed the project. Furthermore the 3 oz wedge project has clearly defined objectives (*ie* to finalise the product specification, organise sampling during the production run, have the laboratory carry out analysis on the samples and prepare a quality control report for the client). It is obvious that the task can be delegated in its entirety to Barry Evans.

Ralph Peters did not feel he could leave the conference as he still had to present an important paper on the food industry's possible response to the new legislation. He therefore called Barry by telephone and carefully reviewed the project with him. (It should be recognised that this method of communication is significantly less effective than a face-to-face meeting which should have occurred before Ralph's departure.) Having established that Barry felt the project was one he could handle, providing he was given some input from the plant manager on the technical aspects of the manufacturing process, Ralph instructed him to proceed with the work. Finally Ralph emphasised that if Barry ran into any problems, he should not hesitate to contact him immediately.

Upon Ralph's return he was pleased to find that the project had proceeded very smoothly. He felt that there were a couple of details in the final report which might need changing and these he discussed with Barry. Having gained Barry's confirmation that the samples were in the process of being shipped, he called Charles David to (a) let him know the product was on its way to the customer and (b) communicate the contribution made by Barry Evans.

Subsequent problems

Some months later, Ralph Peters and the plant manager were discussing the production of the 3 oz wedge. The client had established a very narrow tolerance on the fish flesh/breading ratio. Due to normal manufacturing variables such as settings on the fish block chopper, viscosity of batters, crumb size variations in the breadings and temperature variations in the freezer, the quality control production supervisors were being forced to reject a very high proportion of the product during any one shift.

Ralph agreed with the production department's position that one possible solution was to change the quality control sampling procedures. The production staff could then be more rapidly alerted to a developing problem early in the production run. In this way it would be possible to make adjustments in the manufacturing process and thereby reduce the proportion of output being rejected because it is outside the manufacturing specification.

Subsequent discussion with the quality control line supervisors indicated that they felt it would be feasible to change the sampling procedure. The problem which they could foresee, however, was that the laboratory analysts would not provide the results of their analysis in time to implement a faster recommendation on modifying the manufacturing process. The analysts, on the other hand, felt that they could produce the results more rapidly, but had no confidence that the line supervisors would either deliver the samples on time or bring any variance in specification to the immediate attention of the production department.

Ralph knew that the only way to resolve the problem was to call a meeting of his department and to discuss the issues with the objective of achieving a consensus over new procedures. From previous experience, he was unwilling to do this because for some reason his meetings always seem to dissolve into an argument. Hence his usual approach was to mandate any changes that were required, knowing that he would then have to endure a few weeks of complaints from his subordinates until they had accepted the revised procedure.

Groups and meetings

In an increasingly complex world, it is virtually impossible for individuals in an organisation to cope with problems and implement decisions by working on their own. Managers are forced to work in and through groups in order to exploit the benefits that can be gained by drawing upon all the expertise available within an organisation.

Another reason for the use of a group-based approach to the work process, is the growing acceptance of the concept that organisations can gain significant benefits from a more participative approach to management. Involving staff in key decisions can increase the probability that these decisions will not be resisted by employees when, for example, they represent a revision in operating procedures.

A significant element of the process of group management will involve informal meetings or more formal committees. Managers are frequently very derisive about committees, making such comments as 'a camel is a horse designed by a committee'. In reality the committee structure can be a very effective decision-making system. For this to occur, however, participants must be trained in the skills needed to optimise the decision process in this type of environment.

Given the natural aversion to group activity by the average manager, extensive research has been carried out on the decision-making process by employees acting alone or in the group situation. It would appear that at the idea generation stage, most people are usually more creative when acting alone. The group environment, even when specific actions are taken to remove inhibitions among participants (eg brainstorming sessions), usually causes individuals to be unwilling to voice all of their thoughts for fear of being criticised. It would seem logical to assume that idea evaluation would also be inhibited by a group situation. However, what seems to occur is that the interactive support of the group members causes a shift in individuals' attitudes towards being prepared to take greater risks and adopt a more creative solution to a problem. One possible reason for this is that individuals link aspects of each other's ideas together and in doing so, establish a more synergistic approach to problem solving.

Researchers have established that groups behave differently due to a number of factors. It should be recognised, however, that there are significant variations in opinions between management theorists on which factors will ensure good or bad performance within a group. Some researchers believe a heterogeneous group in terms of personality, attitudes and experience will lead to the most productive solutions. Nevertheless evidence also exists that this type of group may fail to reach agreement and/or reach the wrong conclusions due to wide variation in opinions. It can probably be concluded that it is group cohesiveness, more than homogeneous versus heterogeneous composition, which is important in ensuring a high quality decision-making process. Where a group has developed a 'team spirit', the feelings of unity within the group will mean that participants spend less time in dispute and can concentrate their efforts on solving the problem confronting them.

Group size can also be an important influence on performance. Large groups (*ie* more than ten members) will generate more ideas, but as group size increases the interactions of such a large number of potential contributors may interfere with reaching a final decision. A group of only two to four people can make participants anxious about their personal visibility and reduce their willingness to contribute to the discussions. In general, therefore, it seems a group size of five to ten individuals will result in an adequate review of the issues and still retain an ability to reach a consensus conclusion on recommended next actions within a reasonable time frame.

Effective meetings

Ultimately, the factor which most influences the group decision process is the behaviour of the individuals who make up the group. The essential component is that there are sufficient people present who can contribute the knowledge and experience necessary to solve the task under review. The other important element is that participants exchange information, ask questions and contribute opinions in a way which creates a climate of effective co-operation.

Professor Maier has recommended a number of principles to ensure a group is effective. These include:

- Effort being directed to overcoming surmountable obstacles, while not creating new obstacles.
- Available facts should be used even when they are incomplete.

- The starting point of the problem is the richest source of potential solutions.
- Disagreement can lead to either bad feelings or innovation depending upon the chairman's abilities.
- Idea generation should be kept separate from the evaluation phase of the meeting.
- The chairman should avoid presenting his own ideas or evaluating any proposals, as this can result in the meeting adopting a certain solution because the chairman is seen to favour the idea.

As participants in a meeting, managers and subordinates should acquire the skill of acting constructively. This involves the capability of seeking more information from others instead of disagreeing with them, and even when one does disagree, seeing that elements of the other person's proposal can be incorporated into the solution. A very common aspect of a successful meeting is that the participants have successfully linked each others' contributions together in the definition of an effective solution. In this way the quality of the solution is optimised; and even more importantly, because the majority of participants probably contributed one or more ideas, the group is usually more willing to implement the proposed next actions.

Effective chairing of meetings

Chairing a meeting is possibly one of the more difficult management tasks. For most managers it is a role in which they are frequently involved as the head of an operating unit within the organisation, but one for which they may have received little formal training.

Effective chairing requires careful attention to planning prior to convening the meeting. First, the manager should determine what the meeting is supposed to achieve and at what point (*ie* during or after the meeting) a final decision will have to be reached. Depending upon the knowledge-skills required to examine the problem to be discussed, and the nature of the decisions to be made, the manager can decide who to invite or not to invite. Where there is an opportunity to involve subordinates in the meeting who would gain from the experience, even if they lack the skills necessary to contribute, this often proves useful to the further development of the employees. Finally the manager should draw up an agenda to be circulated to participants. Where possible this should be accompanied by (a) any information which the manager wants reviewed before the meeting and (b) a request that participants bring additional data or prepare reports which might be required during the meeting.

Once a meeting starts, the role of the chairman is not to express opinions but to work with the group to fulfil the objectives of the meeting, maximise the creativity of contributions, and direct criticism along constructive, not destructive, paths. There is a tendency for most meetings to go on far too long. The chairman should manage the meeting by ensuring that adequate time is allocated to each agenda item, while at the same time endeavouring to reach a consensus on each issue in the time available. It is important for the chairman to summarise decisions or agree next actions before moving onto the next item on the agenda. In this way, no-one is left with the feeling that an issue has been left unresolved. The other advantage of a clear statement of action is that it makes it more difficult for a group member to return to an earlier point in the debate and waste time by re-opening the discussion.

At the end of the meeting, the chairman should briefly summarise all the decisions reached and specify any actions that individual group members have agreed to undertake. Where appropriate, the chairman may wish to circulate a summary (or minutes) of the meeting to participants as an agreed record of events. This summary can then be referred to at a later date and/or be used in a subsequent meeting to clarify any misunderstandings.

Avon Foods – effective meetings

Ralph Peters decided that the time had come to determine why meetings of the Quality Control Department were less than productive. He decided to canvass the opinions of his subordinates. Areas of common complaint included:

- The purpose of many meetings did not seem very clear.
- People often arrived completely unprepared.
- Meetings seemed to go on too long.
- Ralph was seen to dominate the discussions, pushing his ideas and criticising everybody else.
- Problems were rarely carefully analysed before people started to propose solutions.
- Everybody was too critical of others and not of themselves.
- Ralph frequently modified decisions outside of the meeting.

Having considered these findings, Ralph scheduled a meeting at the end of the week and issued an agenda which clearly stated that the only topic was to attempt to solve the problem of reducing the level of rejection for the 3 oz wedge product. He attached to the agenda a summary of the current procedure, which is to take a sample of 30 wedges every two hours (*ie* at 10 a.m., noon, 2 p.m. and 4 p.m.) The laboratory is then required to produce a full report to be ready when the next samples arrived (*ie* at noon, 2 p.m. and 4 p.m.) and at the end of the shift. All staff were requested to review these procedures prior to the meeting and Ralph suggested that they may wish to discuss their ideas with each other before the end of the week.

At the beginning of the meeting, Ralph made it clear that he hoped to fulfil the role of a chairman, not decision maker. He asked that the group immediately tell him if at any point during the meeting they felt he was attempting to influence the discussions. Ralph then briefly reviewed the current procedures and asked the group to express their opinions about the various aspects of the problem without criticising the previous speaker. After 20 minutes, he summarised the various comments on a flip chart and requested that the group now begin to use the summary as the basis for constructing a possible solution.

The consensus was that (a) the period of two hours between samples was too long and (b) the number of wedges analysed per sample was too large to be handled in a short time. The line supervisors proposed that sample frequency should be reduced to once an hour and the analysts were confident that only five wedges needed to be examined to determine if there was any variance from specification. They also felt that by rescheduling of other tasks on the days that the 3 oz wedge was in production, they could produce the analysis in 30 minutes. They could then inform the line supervisors of any variance and not await the arrival of the next sample batch. These revisions in procedure were accepted by the group for immediate implementation.

The food technologist Barry Evans also raised the issue that the client's specification was a little too rigid, and that a slightly wider tolerance range would significantly reduce the level of product rejection. He was asked by Ralph to prepare a report on this issue which compared the client's tolerances against normal industry practice. This would then provide Ralph with the necessary information to raise the matter with the client.

The meeting, much to everyone's surprise, was over in 54 minutes. Ralph therefore requested permission to raise a non-agenda item: namely whether the new approach to meetings was seen as more effective. He was pleased to discover that people felt Ralph had performed well as a chairman, instead of decision forcer or influencer, and that there had been a very constructive atmosphere which led to a better quality of decision being made.

6 Performance appraisal

Case: Scot Banes Ltd

Scot Banes markets bulk pack frozen fish fillets and breaded fish products. The company does not own any catching operations, so both the fish fillets and the fish blocks used in the breading operation are purchased on world markets. Raw material procurement was originally the responsibility of the Production Director, Tom Burnham. Because of an increasing workload, he appointed Graham Jones 18 months ago to the post of Procurement Manager.

Once any member of Tom Burnham's group proved their worth, he let them operate with minimal supervision. At the beginning of the second year of Graham's appointment, having developed confidence in his ability, Tom outlined the annual target for the quantity of fish to be procured. He also then made it clear that Graham was now completely responsible for the day-to-day operation of fish procurement for Scot Banes Ltd.

In order to avoid any misunderstandings about Graham's role, Tom prepared a detailed description of the Procurement Management's job. Key elements of this job description were as follows:

Title: Procurement Manager
Accountable To: Production Director

A. Purpose of Job

– Procurement of fish fillets to meet the needs defined by the marketing department sales forecast, and fish blocks to fulfil the raw material requirements of the company's breading operation.
– Provision of information on raw material prices to the marketing department on a timely basis to assist their pricing decisions.
– Assisting the Production Director in (a) determining procurement

activities relative to the company's annual plan, and (b) revising procurement practices in relation to new sales forecasts and/or finished goods inventory balances.

B. Main Areas of Responsibility

- Negotiation and procurement of fish fillets and fish blocks.
- Preparation of reports for the marketing department on current prices being paid for raw materials and providing guidance on possible changes in raw material availability.
- Approval of all invoices from suppliers prior to payment by the accounting department.
- Co-ordination of fish block procurement activities with the production schedules issued by the Production Director.
- Assisting the Production Director in the preparation of annual procurement plans, establishing procurement budgets and where necessary during the year, revising procurement budgets in response to changes in sales forecasts and/or finished goods inventories.

Tom monitored Graham's activities over the next few months by meeting with him on a regular basis, and also noting any comments by members of the marketing, accounting and production departments. It soon became apparent that Graham was not exhibiting the level of initiative, enthusiasm or self-responsibility which had originally caused Tom to give him a greater degree of job role independence.

Tom decided to discuss the problem with Scot Banes' recently appointed Personnel Director, David Stark. David Stark read Graham's job description and listened to Tom's description of Graham's current performance. David's initial response was thus:

'The job description tells Graham what the job of Procurement Manager entails, but from what you have told me I suspect that Graham has no way of knowing how well he is performing in his new, more independent role. If this is the case, I am not surprised to hear that he is showing a lack of confidence. You need to consider the creation of a more formalised approach to performance appraisal, so that you and your subordinates have a better understanding of how well each of them is doing in their jobs.'

Employee performance evaluation

The more traditional approach to employee performance evaluation is to use a judgement-based system. This method usually involves the manager in completing a series of rating scales or open-ended statements about various aspects of an employee's job role. A common format is illustrated by:-

(1) 'Rate the employee on how well the individual responds to direction.'

1	2	3	4	5
Very poorly		Average		Very well

(2) 'How much supervision does this individual require?'

1	2	3	4	5
Great deal		Average		Very little

The completed questionnaire is then reviewed with the employee. It is at this point in the methodology that any potential problems will become apparent. The manager and the subordinate both realise that the analysis is subjective and incomplete. This situation is further compounded because most managers will tend to avoid extremely critical ratings. Finally the subordinate, in seeking to avoid confrontation, will limit his response to those areas in which he has received a good or excellent rating.

Having identified the weakness in this judgement-based type of evaluation of employee performance, behavioural scientists have been seeking alternatives over the last three decades. Their research has established that effective employee evaluation can only be achieved through equal participation by both the manager and the subordinate. The two individuals should jointly seek to work within a system which permits:

- Analysis of the job role to determine the objectives to be achieved by the subordinate over a specified time period.
- Measurement of actual achievement.
- Analysis of the variance between actual and planned achievement.
- Resolution of how performance can be altered in order to minimise performance variance in the future.

Setting and managing by objectives

For any individual to perform effectively, they must have a sense of direction against which to measure their progress. This can be provided through the setting of objectives for the employee. These objectives must have the characteristics of being specific, realistic, measurable and achievable over a defined period of time. If the organisational objectives can be framed in a way which fulfils these four criteria, research has shown that this type of performance target can have an important influence on self-motivating the subordinate to improve their performance.

Recognition of this capability to improve the performance of employees through establishing objectives which offer positive motivation has provided the basis of a management system known as 'Management By Objectives' (or 'MBO'). The key steps in the process are:

- The manager provides the subordinate with a framework related to the purposes and objectives of their job role. One approach, for example, is to utilise selected elements from the job description and the departmental annual plan.
- The subordinate proposes a list of objectives that he feels can be achieved over a specified time period. These are discussed with the manager, sometimes modified, and then agreed by both parties.
- The subordinate reviews his own progress against the objectives and discusses his conclusions periodically with the manager.
- The sequence is repeated at the end of each specified time period, and where appropriate, new or revised objectives are mutually agreed.

MBO was initially embraced with enthusiasm by many major corporations because (a) it represented a significant advance over judgement-based approaches, and (b) it was seen as being compatible with the concept of a more participative approach to management. But by the mid-'70s, doubts were beginning to be expressed about MBO.

There is a problem with the MBO approach when it is introduced into an organisation where the management philosophy remains authoritarian. The participative aspect of MBO under these circumstances merely contributes further to the tensions already present between managers who are seeking to retain absolute control, and subordinates who wish to have a greater involvement in the organisation's decision-making. Another problem with MBO is the amount of time required of the manager to install and operate the system, although this applies to any large-scale employee evaluation system and is not unique to MBO.

But the area of greatest concern about MBO is its potential to over-emphasise the achievement of short term goals. This emphasis can be to the detriment of identifying and developing ways of improving employee performance over the longer term. There is also a problem when attempting to establish appropriate, measurable objectives for senior managers. Senior managers are frequently involved in activities of a very diverse nature, and it is often difficult to specify objectives which have a direct relationship to any single area of performance such as sales, costs or profits.

Behaviour-based evaluation

In order to overcome the criticisms of MBO, management theorists have proposed alternatives which shift the emphasis on appraisal towards giving subordinates a better understanding of how they are carrying out their job relative to what the company considers to be 'good management practice'. This type of performance analysis is based upon rating actual behaviour at work against previously defined standards.

The approach centres upon collecting information on the behaviour of employees in actual situations, and then using these data to construct scales which describe what is considered to be typical behaviour for levels of performance ranging from excellent to poor. For example, one dimension of an assistant credit controller's job might be record keeping on the status of delinquent accounts. 'Extremely good performance' would be to have up-to-date records on all accounts listing amounts owing, age of debts and any information received from credit bureaux.

In contrast, 'extremely poor performance' would be to have more than 90% of account records not up-to-date, no analysis on age of debts and no recent information from credit bureaux yet entered in the files. Levels of performance between extremely good and poor would be based upon the proportion of account records that are up-to-date, some analysis on age of debts and information entered from credit bureaux.

Behaviour-based appraisal systems tend to be somewhat qualitative in the measurement criteria used. Under these circumstances, unlike the very quantitative MBO approach, subordinates may have difficulty in fully understanding their performance relative to the various parameters against which they will be evaluated.

Integrated objective and behaviour appraisal

Given that MBO provides the subordinate with a measurement of what should be achieved, and a behaviour-based system, and understanding of how to achieve managerial excellence, many organisations are now adopting appraisal techniques which incorporate both forms of evaluation into their employee assessment.

As illustrated in *Fig 6.1*, by assessing performance against the two parameters of achieving objectives and managerial behaviour, employees can obtain a much greater insight on how to enhance their over-all performance. Ultimately the manager hopes that all employees will be located in Cell 1 of achieving a high rating for attaining specified objectives

64

	CELL 1	CELL 3
HIGH	High achievement of objectives. High standard of managerial behaviour.	Low achievement of objectives. High standard of managerial behaviour.
LOW	CELL 2 High achievement of objectives. Low standard of managerial behaviour.	CELL 4 Low achievement of objectives. Low standard of managerial behaviour.
	HIGH	LOW

QUALITY OF MANAGERIAL BEHAVIOUR

Fig 6.1   Performance appraisal grid

and at the same time exhibiting a high standard of managerial behaviour.

To demonstrate the integrated evaluation approach, the system can be applied to the appraisal of a salesman employed by a company producing IQF and battered scallops. The company has also recently launched a scallop-based seafood entree. The salesman has agreed with the sales director on achievement objectives related to a specified sales target for existing products and a certain level of distribution in his accounts for the new entree product. Effective management behaviour requires completion of sales reports, handling customer complaints and credit problems, organising product usage seminars for customers, and liaising with the marketing department on the generation of market data to assist in planning future company growth.

A salesman who performs well along both dimensions would be located in Cell 1 of *Fig 6.1*. The manager may then wish to discuss new objectives or aspects of management behaviour related to further developing the salesman to the point where he could be promoted to a more senior position.

A salesman located in Cell 2 will have achieved the required sales target and gained adequate distribution for the new product. However, failure to carry out the management aspects of the job role means that he is probably storing up future problems for the company. The reason for this is that customer complaints are not being handled, customers have not been trained properly in the use of the company's products, and the marketing department does not have the data needed for effective long term planning.

In Cell 3, the salesman has carried out all the tasks which will keep the customer satisfied and the marketing department will have the information it needs. The problem here is that the company faces a financial problem because the salesman has failed to deliver the required sales or gained distribution for the new product.

Obviously, the worst outcome for both the manager and the salesman is a rating located in Cell 4. The individual in this case has neither delivered the quantitative objectives on sales nor executed the various tasks related to managing the customers.

For individuals located in Cell 2, 3 or 4, hopefully the manager and the employee can work together to improve performance over the longer term. Poor attainment of objectives may be due to environmental variables (*eg* unexpected high level of competition; over-optimistic sales forecasts by the marketing department) over which the salesman will have little control. If the poor performance is due to the employee's lack of ability to sell, the sales director would ideally arrange for him to attend an appropriate selling skills training programme. Poor performance along the management behaviour dimension reflects a very common tendency among sales people: namely excessive emphasis on generating sales to the detriment of any other activities. This can usually be overcome by a training programme to make the salesman more aware of the importance of such issues as customer relations, customer education, and the generation of market data for use in long-term planning.

Establishing an integrated appraisal system

The establishment of an integrated appraisal system is not only time consuming but also demands a high level of commitment from both the manager and his subordinates. For each objective, a system must be developed which quantitatively measures the subordinate's progress towards fulfilling that objective. This system must use information generated during the subordinate's execution of the specified task being evaluated. Concurrently a behavioural description will have to be created which defines how the subordinate should carry out certain tasks in accomplishing the stated objectives. It is absolutely crucial that this process should be based upon input from both the manager and the subordinate through a series of discussions, debates and negotiations.

The outcome of all of these activities is the development of performance appraisal forms which contain information on:

- Stated objectives
- Actual performance
- Definition of standards of behaviour from excellent through to poor or inadequate.

Once the appraisal system is in place, the manager and the subordinate can use the analysis as the basis for reaching mutual agreement on current performance, and more importantly, what actions are considered necessary to enhance the subordinate's achievement of objectives and level of managerial behaviour in the future.

Scot Banes – performance appraisal

It is recommended practice to avoid establishing too many objectives for the subordinate. This is especially true during the early phase of a new appraisal system, because the employee can be overwhelmed by what appears to be an impossible range of tasks. Hence when Tom Barnham met Graham Jones, they decided to concentrate on the following key objectives:

- To produce the 20% increase in the quantity of fish blocks forecasted for the next six months at an expenditure level only 10% higher than the same period in the prior year.
- To produce the 10% forecasted increase in the quantity of fish fillets required in the next six months at an expenditure level only 5% higher than the same period last year.
- To review all invoices from suppliers due for payment within five working days of receiving them from the accounting department.
- To develop and produce a monthly reporting system for the marketing department which compares prices paid with both budget and prior year.
- To develop and produce a system which provides the Plant Manager with a weekly update on fish block inventories relative to planned production schedules and where applicable, highlights any potential raw material availability problems.

For each of these five objectives, a behaviour rating scale was then developed. Given Graham's relative inexperience, these scales were initially designed by Tom Barnham working in conjunction with the Personnel

Director. Having produced the preliminary specifications, Tom ensured that they were both understandable and acceptable to Graham. An integrated objective and appraisal form was then created for each of the five objectives.

To illustrate the new system, detailed below is the form which covers objective 1, fish block procurement:

SCOT BANES PRODUCTION DEPARTMENT APPRAISAL FORM

Job title:	Procurement Manager	Name: Graham Jones
Reviewer:	Production Director	Name: Tom Barnham
Review period:	6 months	Review date: XXXX
Job objective:	Fish Block Procurement	

	Budget	Actual	Variance vs. Budget	Actual Prior Year
Quantity				
Expenditure				
Comment on attainment of objectives				

Managerial behaviour appraisal

Instructions for reviewer: The following scale provides guidance on aspects of behaviour ranging from excellent to poor. Based upon your observations of typical performance, please provide a rating for the individual's actual performance over the review period.

Level of performance	*Score*	*Components of behaviour for this specified level of performance*
Extremely good	10	Complete knowledge of price, quality and availability of offerings from all major suppliers on world markets. Has up-to-date information on current and expected catch statistics for all major fishing grounds. Well briefed on current and expected relationships of all major world currencies.
Good	8	Well informed on price, quality and availability of offerings from most

		major suppliers on world markets. Has up-to-date information on catch statistics for most major fishing grounds. Working knowledge of world currency trends.
Average	6	Reasonable knowledge of price, quality and availability of most major suppliers. Some information on catch statistics. Knowledgeable on US/Japanese Yen relationship.
Poor	4	Limited knowledge of price, quality and availability quotes from some major suppliers. Limited knowledge of catch statistics and little understanding of currency trends.
Extremely poor	2	Knowledge of price, quality and availability quotes restricted to 2 or 3 major suppliers. No information on catch statistics or world currency trends.
Reviewer's rating	–	

Role clarification analysis
How could the individual improve his performance in the area of fish block procurement over the next six months?

Potential problems in performance appraisal reviews

It is very easy for the performance appraisal review to run into problems. To minimise these, it is vital that both the manager and the subordinate recognise that the system is designed to provide the basis for further development of the employee in the future. Both parties should therefore be committed to understanding the system and applying the findings to decisions over promotion, training and salary administration.

The core of the performance review is the effective communication between the participants. If either lack these skills, training should be provided prior to implementing the review. Ultimately the factor which will influence subordinate's acceptance of the system is their satisfaction that it provides a way for them to progress in the organisation. At the end of the review session, therefore, the manager and the subordinate must prepare a plan of future actions designed to assist the subordinate to overcome areas of skills weakness. Furthermore the appraisal system must be perceived by all subordinates as an integral component of the management culture of the organisation, providing analysis and feedback on a regular basis. Or put another way, performance appraisal systems are designed to improve performance – not to create more paper to be filed away without action or results.

7 Reacting to non-routine problems

Case: Portmouth Ltd

Portmouth is based in the northwest coastal zone of North America, farming salmon using pen culture techniques. Their output is sold fresh, frozen and smoked to major restaurant chains and large supermarkets.

Mr John Nor, the President, has just returned from a meeting with a major customer who informed him that they are considering obtaining their future supplies of frozen salmon from European sources. The stated reason for this move is that due to a strengthening of the US dollar, European prices are now 15% lower than Portmouth's. Mr Nor has been told he has ten days to review whether he is willing to adjust his company's prices or risk losing future orders.

Problem identification

The starting point for managing any problem is to recognise that a problem exists. Where possible, the manager should anticipate problems before they create an impact on corporate performance. One way to achieve this is for the company to establish a control system which monitors trends both inside and outside the organisation. Justification for the cost of such systems is that the sooner a company recognises the existence of a new problem, the less damage it is likely to inflict on corporate performance.

In addition to standard control systems, managers should also monitor the business environment for informal signs of newly developing problems. Mr Nor, the President of Portmouth, has received information on potential future sales to one customer. More importantly, however, this information is an informal indication that Portmouth faces an even larger problem: namely severe price competition from European suppliers across the entire market. The importance of informal information sources is greatly underrated by many organisations. Having invested heavily in sophisticated monitoring systems using computerised data bases, it is not easy to accept that managers should continue to allocate a significant proportion of their time to gathering important facts through informal conversations with suppliers, competitors and customers.

Handling problems effectively

The ability to deal with problems effectively is a major requirement of managers at every level within an organisation. It requires the following:-

- *Knowledge of the subordinates' skills, coupled with a complete understanding of the relevant objectives, policies and procedures within the organisation.* For example a sales manager might wish to reduce the level of paperwork for his group. One solution could be to abolish the policy that a travel approval form must be submitted by every salesman before they can make a trip to see a customer. He knows that less time spent on filling out standard forms would leave the sales force with more time to execute important administrative tasks (*eg* analysing recent sales of key customers versus budget), and his staff are sufficiently responsible not to abuse the company control systems. Nevertheless the company policy requires that significant expenditure does not occur without prior approval by a senior manager. The sales manager may therefore create a compromise which will fulfil both objectives. This might be to waive the need for prior approval forms on sales trips unless these involve either overseas travel or being away for more than 72 hours.
- *The skills to identify, analyse and develop appropriate solutions. Linked to this, the manager must have the ability to recognise when he should turn to others in the organisation who could help in the problem/solution process.* The sales manager mentioned above, for example, might believe that another way to reduce paperwork is to have his sales staff make much greater use of computers to analyse sales performance by account. As he is aware that he lacks the skills necessary to design and implement this type of solution, he requests the assistance of the company's Data Processing Department.
- *The ability to review information on a specific issue and then judge which is the most appropriate solution from the various alternatives available.* The sales manager in our example may have received a proposal from the Data Processing Department on various options ranging from a simple on-line analysis package using the company's

mainframe computer system through to equipping the sales force with individual desk top computers capable of running sophisticated spreadsheet software systems. The sales manager must then weigh up the cost/benefit implications of the various alternatives and judge which is the most suitable for the task concerned.

Analysis of the problem

Most problems require a certain degree of analysis before any decision can be reached. The depth of analysis will vary, however, depending upon the nature of the problem. If it is of a routine and recurring nature, there is usually an established standard procedure for handling it. Also if the impact of the problem on the operation is minimal (*eg* a new departure time for a trawler will have to be determined because of a delay in the loading of ice), a manager will not bother to spend too much time examining what is the most appropriate decision.

When a non-routine problem occurs, especially if there are indications that it could have a major impact on corporate performance, then a much fuller analysis will be required. It is recommended that the following actions are instigated:

- Gather information about the relevant issues and attempt to distinguish between fact and supposition.
- Establish the level of priority that should be given to resolving the problem, relative to other managerial responsibilities.
- Evaluate the consequences of the problem in terms of potential impact on corporate performance.
- If the first three actions are sufficient justification for further analysis prior to reaching a decision, create an appropriate framework for gaining a fuller understanding of the causes underlying the problem.

Where a problem is very complex in terms of cause or consequent impact, it may prove useful to form some form of diagrammatic representation of the issues as a way of visually summarising potential impact on the company. Further understanding may also be created by accompanying this by charting the causes of the problem.

An analysis of the Portmouth problem

Mr Nor's reaction to the information on new, lower prices being quoted by competition, is that this represents a problem which will have a major impact on the future performance of Portmouth. Hence he felt it should receive urgent attention, utilising specialist input from his immédiate subordinates. Before spreading alarm and despondency, however, he initiated a number of contacts with people in the industry to find out whether the European quotes were genuine, because it could have been that the customer had modified the truth in order to place pressure on Portmouth to reduce prices. Unfortunately the additional market intelligence gathered by Mr Nor served to confirm that the European producers were indeed planning a very aggressive attempt to build business in the American market.

Mr Nor's diagrammatic representation of the problem (*Fig 7.1*) shows that the potential impact of price competition could be disastrous because a response of matching the lower price or doing nothing would in either case reduce profitability. Even if the downturn in profitability was short-lived, it would be sufficient to place extreme pressure on the servicing of Portmouth's existing financial liabilities and on their plans to raise additional funds to support an expansion of the salmon farming operation.

A fuller analysis of the underlying causes of the European price decrease reveals that the strengthening of the US dollar was not the sole cause for the aggressive posture of the European producers (*Fig 7.2*). Other factors include government subsidies, economies of scale, greater productivity and limited alternative market opportunities. Hence Mr Nor is forced to conclude that Portmouth is not facing a temporary situation of just having to wait until there is a realignment of the US dollar against European currencies.

Generating solutions for non-routine problems

When faced with a problem, many managers tend to adopt the first solution which occurs to them. This is probably acceptable for standard, routine problems or if the manager is a natural entrepreneur. Nevertheless this approach blocks any chance of a more innovative solution being considered. In the case of problems which have the potential for a major impact on corporate performance, the failure to review a number of alternatives can result in the acceptance of a sub-optimal solution.

It is important that managers develop a creative approach to the process of generating alternative solutions for non-routine problems. This demands an open mind and a willingness to avoid relying too heavily on past experience, especially if such experience is not relevant to the current situation. During the phase of generating alternative solutions, the manager should attempt to suspend the natural desire to critically evaluate each idea until a number of scenarios have been developed. It can sometimes be

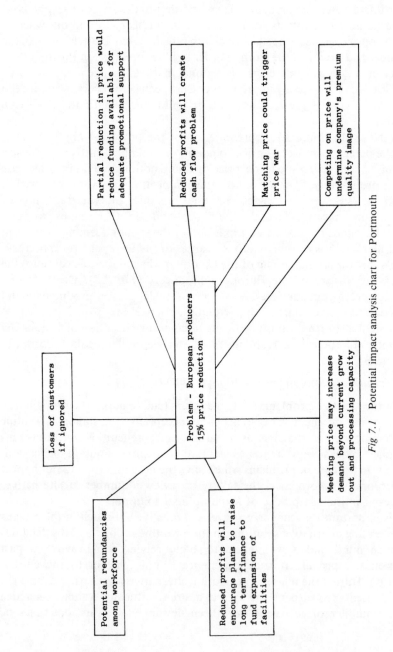

Fig 7.1 Potential impact analysis chart for Portmouth

The boxes in the chart contain the following text:

- Partial reduction in price would reduce funding available for adequate promotional support
- Reduced profits will create cash flow problem
- Matching price could trigger price war
- Competing on price will undermine company's premium quality image
- Loss of customers if ignored
- Problem - European producers 15% price reduction
- Potential redundancies among workforce
- Reduced profits will encourage plans to raise long term finance to fund expansion of facilities
- Meeting price may increase demand beyond current grow out and processing capacity

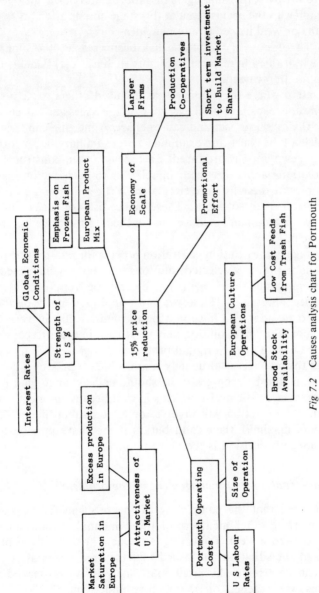

Fig 7.2 Causes analysis chart for Portmouth

beneficial to involve other managers or subordinates in the idea generation process, again with the proviso that at this stage no-one attempts to evaluate the strengths or weaknesses of any new idea.

Once a list of possible solutions has been created, the manager can evaluate each of them in relation to factors such as Cost, Benefit, Risk and Time-scale for Implementation.

In a situation where the manager has involved others to help solve the problem one approach is to brief everyone on the issues and ask each person to produce their own impact and causes chart. A meeting can then be held during which all the charts are examined and a summary set, incorporating the varying viewpoints, is produced. Each person then constructs a list of possible solutions and rates them in relation to such factors as cost or benefit. Copies of these lists are circulated to the group and each member selects their top three solutions. These 'short lists' of ideas can then be discussed by the group and a consensus reached on the three best solutions to the problem.

The last stage in the solution generation process for non-routine problems is to re-examine the three selected solutions to see if any improvement can be made by combining them together or by drawing from other ideas which may have been presented. Finally the group leader must decide which solution is to be adopted. It should be recognised that in very complex situations, more than one solution may be utilised in order to cope with the multi-faceted nature of the original problem.

Although the manager is ultimately responsible for making a decision on which solution is to be progressed, he should endeavour to gain his whole team's acceptance of the decision. This is vital because in most situations, implementing the solution will involve, or affect, subordinates. They are more likely to maximise their contribution if they have previously agreed that the manager's decision is the best option.

Portmouth – analysis and management of the problem

Mr Nor decided that the complexity of the problem facing Portmouth required involving other members of the company's senior management team. He therefore held a meeting at which he briefed them on his perception of the situation and asked that they reconvene after everyone had had a chance to review his preliminary analysis. At this second meeting, a consensus was reached on the factors underlying the level of price competition from European producers. It was decided that the proportionate influence of the various factors contributing to this price reduction could be broken down as follows:

4.5%   caused by the strengthening US dollar.

3.0%   short term promotional discounts to gain customers which would probably only be sustained for a few months.

3.0%   reflecting the emphasis on the marketing of lower quality, frozen product.

3.0%   reflecting the location of the farms in areas of Europe benefitting from government subsidies and grants to stimulate employment.

1.5%   due to the economy of scale of the European operations compared to Portmouth's scale of production.

15.0%   total level of price competition.

The Portmouth team's solutions for reacting to the price competition problem covered both immediate and longer term responses. The most immediate possible reaction would be for Portmouth to announce a 6% promotional discount. This level of discount was all the group felt could be afforded without a significant impact on profits. Arguments against this idea were based upon the facts that (a) the reduction was significantly less than the European price reduction of 15%, and (b) this would signal to the market that Portmouth was willing to enter into price competition. Eventually it was agreed that Portmouth needed to seek a market position which removed the company from competing in more price sensitive areas of the market. It was felt that this could be achieved by overtly positioning Portmouth as a premium quality producer through directing the majority of future promotional support behind the company's fresh salmon. Currently the company only markets a small proportion of production in the smoked form because of limitations in smoking capacity. Market acceptance of the smoked product sold to end user outlets and through mail order had been very high. It was felt that given this strong performance and the higher profit margin available from the value-added product, plans to expand the company's fish smoking production capacity should be accelerated. This product would further assist the objective of positioning the company as the producer of premium quality fish.

It was generally agreed that Portmouth's reliance on a single species did place the company in a vulnerable position if the level of overseas competition further intensified. Another idea was therefore to broaden the product range through the acquisition of another company involved in the processing of other fish species. It might also be possible to protect the company's market position through vertical integration involving acquisition or merger with a seafood restaurant chain or a seafood wholesaler group.

An alternative to these market-orientated strategies was suggested by the managers responsible for farming and processing. They felt that consideration should be given to actions that could stabilise or reduce production costs by accelerating plans for expansion of the farming operation. Other managers were not convinced that this approach offered cost reductions sufficient to improve the company's position versus overseas competition. Hence there was discussion of the concept 'if you cannot beat them, join them', by establishing or buying a farming operation based in Europe. Finally a minority fall back strategy was proposed: namely, if no real protection against overseas competition could be identified, consideration should be given to selling Portmouth.

The group then rated each idea in relation to the factors of Cost, Benefit, Risk, and Time-scale for Implementation (*Fig 7.3*), and used this analysis to rank the possible alternative strategies available to the company. It was not felt that any single action would provide complete protection from overseas price competition, so Mr Nor proposed that Portmouth should implement the top three actions of:

- Greater emphasis on a premium quality market position using fresh salmon as the lead item to receive promotional support.
- Immediate scaling up of production capability for smoked salmon and a more aggressive promotional stance on this product once the company was in a position to supply additional product.
- A move to a broader product line through the acquisition of a company involved in the processing of other premium quality fish species.

From decision to planning

Reaching the point at which a decision has been made usually takes much less time than the processes associated with planning and implementing the selected solution. Once a decision on reacting to a non-routine problem has been reached, the manager will then have to ensure that appropriate plans are developed to implement the decision.

Moving to the plan development phase may require input from subordinates who have previously not been involved in considering an appropriate response to the original problem. Under these circumstances, the

POSSIBLE ACTIONS BY PORTMOUTH	COST	BENEFIT	RISK	TIME SCALE	FINAL RANKING OF POTENTIAL ACTIONS
Acquire or merge with wholesaler or restaurant chain	High	High	Medium	Long	5
Acquire European farm operation	High	High	High	Long	6
React with 6% price reduction	High	High	High	Short	4
Move towards emphasis on fresh salmon in sales mix and overall premium quality position in market	Low	Medium	Low	Short	1
Expand production capacity for value-added smoked fish	Medium	High	Medium	Medium	2
Broader product line by acquiring fresh fish processing operation	Medium	Medium	Medium	Medium	3
Accelerate plans to scale up grow out operations to reduce culture costs	High	Medium	High	Long	7
Sell operation	High	High	High	Long	8

Fig 7.3   Review chart to identify most feasible actions by Portmouth

manager or his representative will have to provide these newly involved subordinates with a briefing on how the decision has been reached.

The manager should also develop a statement of the specific goals to be achieved during the implementation of the decision. These goals will provide the framework upon which plans can be built. As the subordinates will undoubtedly have questions about what resources will be made available (*eg* personnel, equipment and financial), the manager should specify any resource constraints which should be borne in mind during the planning phase. Finally the manager should inform the subordinates of the proposed time-scale for the planning and implementation stages of the new programme.

Where a decision has been made at a senior level within an organisation, it will not be detailed enough to be immediately translated into specific departmental goals. Senior management will, therefore, have to involve their immediate subordinates in further analysis of the problem facing the company as it relates to reaching appropriate departmental level decisions and objectives. Only after these have been agreed can the subordinates move forward and start work on planning departmental-level reactions.

Planning at Portmouth

The marketing department will have to make additional decisions on matters such as which sector of the market offers opportunity for incremental fresh salmon sales, how to handle the de-emphasis of frozen product with existing customers, and whether current promotional resources are adequate to handle the positioning of the company as a premium quality supplier. The production department will face the task of reaching decisions on issues such as the best way to expand fish smoking capacity and the processing of larger quantities of fresh product.

The decision to acquire another company to expand Portmouth's base of products is of such significance that Mr Nor will probably wish to retain control over this project. Nevertheless he is likely to involve his Finance Director because jointly they will have to reach decisions on matters such as the size of company to be acquired and the possible funding alternatives available for this purchase.

As will be demonstrated in chapter 8, compared to the implementation phase, decision making is often the easiest part of reacting to non-routine problems. This fact is only true, however, if the correct decision is made in the first place. In the event that a sub-optimal or incorrect decision is made, this will become very evident as the company discovers that the new or revised plans do not work. Managers must endeavour to maximise the quality of their decisions if an organisation is to continue to exist and grow over the long term.

8 Implementing decisions through planning and control

Case: Aquapron Ltd

Aquapron is a diversified food manufacturing company which owns a shrimp processing operation as part of its corporate portfolio. This subsidiary is supplied with shrimp caught by its own boats and by independent skippers who work on contract. Recent trends in catch have been interpreted by the Government as an early sign of overfishing. It is very likely that the Government's solution will be the introduction of quotas which would reduce Aquapron's source of raw material within two to three years. In the light of this prediction, the Board of Aquapron was very attracted by a Government proposal to make development grants available to companies willing to work with the agricultural community in the joint establishment of aquaculture schemes to raise the Malaysian prawn *Macrobrachium*.

The Aquapron Board therefore instructed the director in charge of the shrimp processing operation, Andrea Ortiz, to progress the establishment of a *Macrobrachium* farming operation to compensate for the expected shortage in marine shrimp landings.

Establishing objectives

Before a manager or his subordinates can develop a plan, it is necessary to define appropriate objectives (or goals) to be achieved. In the case of a complex project involving a number of departments within the organisation, a hierarchy of departmental objectives will also have to be established (*Fig 8.1*). These sub-component objectives when combined will represent the overall purpose of the programme being planned.

The importance of defining clear objectives at the outset of a project cannot be stressed too highly. Research on complex projects in a number of

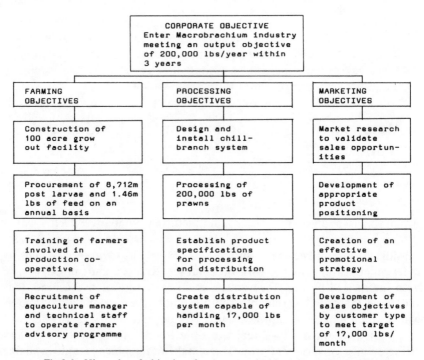

Fig 8.1 Hierarchy of objectives for Aquapron's *Macrobrachium* operation

industries which have encountered difficulties, has revealed that the underlying cause of failure was often that the management team had not reached
agreement on clear objectives and/or not communicated these to their
subordinates. In consequence the managers were endeavouring to create a
plan without having any real sense of direction.

Clearly defined objectives can also provide a benchmark against which to
test the viability of a plan. For example: a marketing manager was given the
objective of relaunching a range of lightly battered shrimp products, but
under the operating constraint that he would not be able to incur any
significant capital costs. When the production department was assigned the
task of manufacturing the new formulation, they found that the improved
product would cause a reduction in throughput on the batter line of almost
20%. As this line was already working at full capacity, the only way to meet
existing sales forecasts and also to produce the new formulation would be to
install an additional batter line. Unfortunately, as no-one had communi-

cated the constraint over capital expenditure, the production department had created a solution which was not acceptable. A clearer definition of objectives by the marketing manager when briefing production staff would have ensured that development work would have concentrated on a solution which did not reduce production throughput capability.

In the event that a plan is found to be incompatible with clearly defined objectives, the project manager faces three possible courses of action, namely:

- Develop a revised plan which meets the objectives.
- Reconsider the correctness of the objectives.
- Terminate the project as impractical.

Aquapron – setting objectives

The rather unspecific nature of the assigned objective from the Board, to 'establish a *Macrobrachium* operation', will necessitate Andrea Ortiz obtaining additional information and/or making some assumptions before a more precise set of objectives can be defined for the project. Wherever possible, quantitative objectives are usually more effective in the consequent planning process than those stated in qualitative terms.

Andrea Ortiz first has to decide on an objective which defines the proposed output for a *Macrobrachium* culture operation. The shrimp plant currently processes a million pounds of marine shrimp per year. The Government Fisheries Department has made a provisional statement that they expect it will be necessary to introduce quotas which will limit catch by approximately 20%. Making the assumption that the *Macrobrachium* would be processed using a standard chill-kill, blanching and icing technique, Andrea estimates that in the face of marine shrimp quotas, the plant could handle up to 200,000 lbs of *Macrobrachium* per annum. This assumption permits him to define an overall objective for the project that 'Aquapron will enter the *Macrobrachium* business with an output target of raising at least 200,000 lbs/year within three years of the project being initiated.'

This overall objective can then provide an umbrella objective under which a hierarchy of sub-objectives can be established for the functional areas of the project which include farming, processing and marketing. These sub-objectives should not be assumed to be completely accurate because there is often an inter-relationship between the sub-objectives. For example, the processing capacity required will be influenced by the expected seasonal pattern of harvesting to be used in the farming operation. In the case of projects that are new to an organisation, further research will frequently be necessary to generate information which is needed before all or some of the sub-objectives can be finalised. Given Aquapron's total lack of experience of *Macrobrachium* culture, it will be this aspect of the project that will demand additional effort to generate further information, and it is very likely that the company will require the services of an external advisor.

This advisor was instructed to develop an outline proposal on the probable operating parameters for a *Macrobrachium* culture system. From this report, Andrea Ortiz was then able to assume that the farm operation would probably use existing technology based upon a pond-based grow-out system, harvested on a year round basis yielding approximately 2,000 lbs of prawns/acre of pond/year. This conclusion will mean that the marketing and processing objectives will be based upon the handling of a non-seasonal product output of approximately 16–17,000 lbs/month (*ie* 200,000 annual production divided by 12).

Developing a plan

The delivery of objectives can only occur after they have been translated into statements of proposed action. Preparing this systematic and detailed description of actions is known as 'planning'. The issues which will have to be covered in a plan are:

- What are the activities to be carried out?
- How do these activities relate to each other and the previously established objectives?
- What resources will be required for each activity?
- How will any new resources be acquired; and if existing resources are to be used, can they be utilised without affecting other areas of the company's operations?
- What are the risks, opportunities and potential problems associated with each activity?
- What is the time frame over which the plan should be carried out?

It is critical that sufficient time is allocated to carry out a plan. One aid to the process is to break down the plan into logical steps, and prepare a flow diagram which links together important sequential actions (*eg* a new factory must be constructed before the new equipment can be installed) and also shows which steps can be concurrent (*eg* building a new freezer while installing a new production line in another area of the plant). Clearly, the more steps that can be structured to run concurrently, the greater the likelihood that the plan will be carried out in the shortest possible time.

Where the plan involves a diversity of specialist sub-objectives, it is very likely that each functional area (*eg* marketing or production) will be required to produce their own very detailed proposal. These proposals can then be combined into a final over-all plan.

The Aquapron plan

Each of the three proposed areas of activity – farming, processing and marketing – will require a detailed plan and flowchart. To illustrate the concept of planning, the text will be restricted to examining the plan for the farming operation.

The elements of the farming plan are summarised in the flowchart shown in *Fig 8.2*. The background to each of the activities mentioned in the flowchart are as follows:

– *The feasibility study* – this will be required because Aquapron has no experience of *Macrobrachium* farming and needs to determine if they should enter this new area of business. The company will contract with the Government's Agriculture and Fisheries Advisory Service to have this study carried out.

– *Formation of the co-operative* – to minimise the need for capital investment, Aquapron had decided to contract out the farming to local land owners whose activities will be co-ordinated through the creation of a 'Producers Co-operative'. The terms of such co-operatives are already specified by Government legislation. Once Aquapron can demonstrate that a new producers co-operative will be established, the company will be able to file a grant application for 100% of the costs involved in pond construction.

– *Project management* – the scale of the operation will mean that Aquapron will have to recruit an experienced scientist to manage the advisory service that will assist the local farmers engaged in raising the *Macrobrachium*.

– *Post-larvae* – to minimise capital expenditure, Aquapron will not establish their own hatchery, but will instead contract with the Government hatchery for a supply of post-larvae.

– *Grow-out facilities* – the intention will be that each farmer will operate, on average, five acres of ponds yielding 10,000 lbs of prawns per year. The basic specification for a pond will be an average size of 2.5 acres with an average water depth of three to four feet and a sluice gate

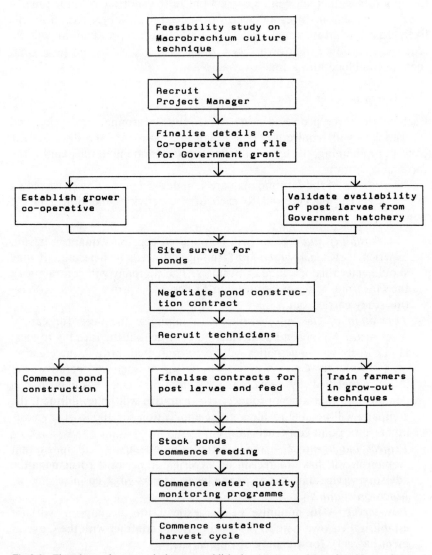

Fig 8.2 Flowchart of proposed plan to establish the Aquapron *Macrobrachium* culture operation

outflow system. The ponds will be fed from a water source providing 15 gallons/minute/acre. The prawns will be fed daily using broiler starter feed which is readily available through local agricultural feed suppliers. The ponds will be stocked once a year in the summer and harvesting will commence once prawn interaction becomes evident. From this point onwards, harvesting will occur every three weeks using a two inch mesh seine net. The catch will be iced before transportation to the processing plant. During the grow-out phase, water quality will be monitored by the Aquapron advisory team for temperature, dissolved O_2, pH, total alkalinity and nutrients.

- *Technical staff and training facilities* – working under the direction of the project manager, a team of technicians will be responsible for the day-to-day liaison with the producers. They will monitor pond conditions and carry out all analyses. These technicians will also be responsible for the training of the farmers in the producers association.

- *Timing and project phases* – although it was originally planned to reach an output of 200,000 lbs of prawns within three years, analysis of required actions indicate that it will take approximately 24 months from the start of the feasibility study to the first harvest. As technical problems can be expected at start-up and during the output expansion phase, it is now planned to take four years to reach the final output objective. The scaling up of production output is now set at 50,000 lbs in Year 2, 100,000 lbs in Year 3 and 200,000 lbs by Year 4.

Controlling a project

It is wise in any organisation to be aware of the possibility that if things can go wrong, they will. The sooner a potential divergence from a plan is identified, the greater the probability that the programme can be brought back on course. This can be achieved by the installation of appropriate control systems which monitor performance and give an early warning of when actual results vary from those specified in the original plan.

The type of control system utilised will vary according to what is to be measured and how it is to be measured. The control system provides a method for recording activities, and these records may be collected by statistical analysis, through a formal report or verbally during regular meetings between the manager and his subordinates.

As with objectives, it is more effective to base the control system on specific data as opposed to qualitative opinion. Furthermore the information will be more meaningful if related back to the standards specified by the original objectives determined at the outset of a project.

Excessive control can be as non-productive as no controls at all if such controls cause subordinates to spend the majority of their time providing information instead of executing their responsibilities. The subordinate also needs to be able to see that the information generated by the control system is being used, not merely filed. Filing information away unused occurs more frequently when organisations establish very large and intrusive controls.

In designing a control system, therefore the manager must isolate those areas of planned activity in which any variance in performance is critical to meeting the overall objectives for the project. Where an area of activity seems relatively straightforward, the manager may be able to omit the need for any control. In addition, once an operation comes on-stream and an area of activity seems to be performing smoothly, it is good practice to simplify or withdraw the relevant control system. This will release subordinates to concentrate on other, potentially more productive tasks.

A decision also has to be made on who is responsible for a control system. Where possible the management of information collection and analysis should be delegated to the lowest possible level in the organisation. In this way, senior managers are able to spend longer time on more creative tasks such as planning and decision making. This delegation should, however, avoid placing the control in the hands of inexperienced or unqualified employees.

Most control systems are 'absolute' because all they can achieve is to indicate that performance is at variance with the original objective. It is more difficult, but certainly more productive, to design a control system which has a diagnostic capability. In this way, the manager is not only aware that a problem of performance variance is emerging, but he will also have information which can provide some degree of explanation of why the problem has arisen. For example, if a company has a profit objective of £100,000 and a report that profits-to-date are £80,000, this is an absolute control. The manager has no details on why profit is less than planned. In contrast, if the control system presented the information in the following form:

	Planned	Actual	Variance
Sales	300,000	300,000	0
Cost of goods	200,000	220,000	+ 20,000
Gross profit	100,000	80,000	− 20,000

then this will indicate that rising cost of goods, not a decline in sales, is the cause of the profit shortfall.

Examples of controls for the Aquapron operation

The company will need control across all aspects of the *Macrobrachium* operation. For illustration of the concept of control, however, the examples will be limited to the farming operation.

During the construction phase the two measurements that should be made are actual time taken for construction versus plan, and actual expenditure on pond construction against budget. Once the culture operation is started, the manager should be very concerned about three key performance variables:

- Number of post-larvae stocked versus final number of shrimp harvested. This is necessary to ensure that mortality during grow-out is minimised.
- Weight of feed per acre versus output per acre. This will provide an indication whether feed to conversion is adequate.
- Actual output per acre versus plan to monitor the productivity of the system.

Given that a number of producers will be involved in the farming activities, it will be necessary to produce information both on an aggregated basis to provide the manager with a complete picture and also a breakdown of performance by farm to isolate any variance in efficiency between producers. Where any variance in feed conversion, post-larvae mortality and output is identified, the manager can then call upon the file of information on water quality and general observations on husbandry that the technical observers have collected during the year.

As the technicians will be in regular contact with the producers, it is appropriate that they are given the responsibility for collecting and analysing data for specific farms. Their individual reports should then be consolidated by the senior technician who will be responsible for producing an aggregate analysis of the entire farming operation.

9 Organisational structures

Case: Place Bay Ltd

Place Bay is a family owned company which has existed for over a 100 years. The organisation operates a trawler fleet and a processing plant. The last significant investment was the refurbishment of the plant in the '60s when the company decided to concentrate on the production of frozen cod blocks used in the manufacture of breaded fish products. In recent years the decline in the cod fishery has caused the company to broaden their processing activities and move into the production of various groundfish fillets packed as 'own label' brands for supermarket chains.

It has always been important to the owners of Place Bay that the company should try to maximise the number of jobs in the plant because it represents virtually the only source of employment in the area. Meeting this social objective – while still remaining profitable – has become increasingly difficult, and the Chairman Mr Theobold became aware that output/employee is well below the average for the fishing industry. He therefore asked an old friend and experienced businessman, Mr James MacBaen, to spend a day looking around the plant.

Mr MacBaen's brief was to provide some new ideas on how to increase productivity with no preconceptions of how a fish plant should be operated.

Mr MacBaen's day began with a detailed briefing from the plant manager on the various activities which go on in the plant and the organisational structure of the operation. A noticeable feature of the briefing was the very detailed information that was provided on daily output, mix of fillet packs, processing yields per pound of round fish and departmental operating costs.

Mr MacBaen spent the rest of the day visiting various departments, observing operations and talking to employees at all levels about their job roles. He was very impressed with the positive attitudes of the employees and their clear understanding of the specific aspect of the processing procedures for which they were responsible. While standing by a cutting table, one of the filleters broke his cutting knife. A supervisor was called over and they discussed the incident in some detail. The supervisor then filled out a multicopy form, which he and the operative signed. Mr MacBaen was intrigued by this procedure and asked about it. The supervisor explained that the purpose of the form was to collect information on the breakage of equipment and also to provide approval for the filleter to draw a new knife from the stores. Before the filleter was able to go to the stores, he would have to also obtain a third signature from the shift supervisor. When this had been done, one copy of the form would be left with the stores clerk, a second would go into the shift supervisor's file and a third to accounting to match against the stock withdrawal slip which the stores completed. A fourth copy would be forwarded to the plant manager's office.

Out on the unloading dock, a catch of ocean perch (or 'redfish') was being unloaded from a trawler. This fish was passed through a mechanical descaler before being stored on ice in holding pens. As Mr MacBaen arrived, the descaling machine stopped working and the dock supervisor explained that it was nothing very important, merely a drive belt which had snapped. This apparently was a regular occurrence. He enquired how long the machine would be out of action and was told that although replacing the belt would take about 20 minutes, the dock supervisor expected he would not be able to restart the unloading for about two hours. The reason for this extended delay was that the specified procedure was:

- The maintenance supervisor would have to inspect the machine to confirm the nature of the fault.
- A repair report would have to be prepared by the supervisor specifying the fault, describing any parts required and estimating the time for the repair.
- The repair report would have to be reviewed by the relevant departmental manager or his deputy who would have to countersign any parts requisition form.
- The maintenance worker allocated to the job would have to draw the parts from the store and obtain a signature on the requisition form for return to the plant manager's office.
- Once the repair was complete, the maintenance supervisor would have to inspect the work and then sign the job sheet which goes back to accounting for them to record the cost of the repair work.

These two events caused Mr MacBaen to pay more attention to procedures when visiting other departments in the plant. It came as little surprise to find that all employees were involved in extensive preparation of forms and records during all aspects of their job process. This situation explained how the plant manager was able to provide such detailed information during the earlier briefing on plant operations.

Classical theories of managing organisations

The traditional theories of organisational management can be grouped into three main categories – scientific management, administrative management and the bureaucratic model. Frederick Taylor is seen as the father of the scientific management school. He was primarily concerned about developing a systematic approach which would permit the standardisation of job roles as the basis for maximising productivity. These concepts provided the foundation for such techniques as time and motion studies and the redesign of an operatives approach to a repetitive task.

Building on scientific management concepts, theorists led by Luther Gulick and Lyndall Urwick proposed universal principles of 'administrative management'. These principles covered such issues as documenting job functions, defining clear lines of management authority, authority channels flowing from the top to the bottom of organisations, individuals having a 'span of control' not greater than five or six subordinates, clear job definitions and finding people to fit these descriptions.

Max Weber, a German sociologist, conceived the bureaucratic model as an organisational structure best suited to the influence of changing social conditions that followed the industrial revolution. Although the term 'bureaucracy' is now synonymous with 'red tape' and ineptness, the original objectives of Weber's proposal was to create the most efficient instrument to operate large-scale enterprises in both the public and private sectors. The key elements of his model were:

- Specialisation of labour, with employees trained to be experts in a specific functional role.
- A hierarchy of authority with each level of management under the control of the next higher level.
- Clearly defined specifications of job role, level of responsibility and accountability.
- Consistency of decision making achieved through reliance on clearly defined policies and procedures.

– Centralisation of authority within the upper levels of the organisation to minimise co-ordination problems between departments.
– Reliance on records and files to preserve information on previous actions so that the precedents could be applied in the future.

Despite the association that people usually make between bureaucracy and Government departments, the characteristics of this type of structure are also to be found in many large private sector companies. The widespread adoption of the concept reflects the fact that it does offer certain key advantages. For instance, viability of the organisation will not depend upon a single executive because the application of well defined policies or procedures mean that any one individual is easily replaceable. Clearly defined policies and procedures also confer the advantage of consistent and uniform decisions. Hence for the production of standardised output over many years, the bureaucratic form of organisational structure is extremely effective.

Unfortunately the forces behind the concept of standardisation of behaviour can be a barrier when there is a need to change in response to different environmental circumstances. In the bureaucratic system employees are rewarded for carrying out specified policies and procedures, not for suggesting new approaches that might be more appropriate in the face of changing circumstances either inside or outside the organisation.

Organic management

Whereas classical management theories focus upon the clear definition of job role and lines of authority, behavioural scientists have for many years supported the approach of a more flexible, humanitarian approach to improving employee performance. Their criticism of classic models is that they ignore the psychological and social pressures which can influence employee behaviour.

The reasoning of the behavioural scientist is based upon observations that bureaucratic organisations may develop some of the following characteristics:

– Employees are orientated to the extreme application of policies and procedures to the point where such rigid adherence can impair productivity.
– The job role becomes so narrowly defined that any individual's performance is impaired because of inadequate personal freedom in the execution of the job tasks.

- The centralised decision processes can cause lower level employees to avoid any responsibility by claiming that they lack the necessary authority to respond to any request for action.
- The hierarchical structure tends to lead to employees adopting a minimum standard of behaviour acceptable to their superiors instead of striving for excellence.
- The highly specialised functional units and the rigidity of policies or procedures means that the organisation may become incapable of any form of response that depends upon a multi-departmental action.

Due to such potential weaknesses, management theorists have proposed that bureaucracies are too 'mechanistic' and that there is a need for greater flexibility within the modern organisation. This reduction in rigidity is known as the 'organic' approach to management and is based upon features such as:

- Much wider definition of job roles with emphasis on employee self-responsibility in the place of the reliance upon rules and regulations to control performance.
- A broadening of employee commitment beyond their own job by being more involved in understanding and delivering over-all organisational objectives.
- Reduction in the level of autocratic management through the delegation of responsibilities to lower level employees.
- Communication biased towards advice or assistance instead of direct orders.
- A much wider span of control with a supervisor responsible for a large number of subordinates (cf the administrative model of five to six subordinates) and a reduction in the number of levels of management.
- Employees willing to take responsibility for solving non-routine problems rather than claiming, 'That issue is outside my area of responsibility' or 'I'm not permitted to make a decision about that'.

As managers have been guided for over 50 years by the bureaucratic or scientific models of organisation, the concept of organic management has not received a universal welcome. Unfortunately some of the early disciples of the organic approach tended to overemphasise the need to permit employees to control their own destinies inside the organisation to the point where some managers saw the approach as a blueprint for organisational anarchy. It is also not unreasonable to suggest that the use of the term 'mechanistic' to describe the more rigid aspects of traditional approaches to management did little to help persuade managers that alternatives might be considered.

Fortunately it is now more widely accepted that the mechanistic and organic models are both extreme definitions. Research has shown that many organisations contain features of both approaches and that they are not mutually exclusive. Departments which operate under stable conditions using standard approaches (eg credit control) will tend towards a bureaucratic structure. In those departments where a more flexible, less routine atmosphere is required (eg research and development), a more organic system will probably be adopted.

Place Bay – revising organisational structures

Revising organisational structures is not an action which should ever be contemplated without a long period of analysis, discussion and review. Hence Mr MacBaen would be unlikely to be in a position to make specific recommendations to Mr Theobold on the basis of a one-day visit to the processing plant. Nevertheless he would be able to identify that the processing operation exhibits the following quite revealing features:

Span of control by supervisors	Very narrow
Number of levels of authority	Very high
Degree of centralised decision making	Very high
Specification of employee job roles	Very narrow

All of the above aspects of the operation indicate that the plant is operated with a very bureaucratic approach. This was probably quite appropriate when the company only produced one product, cod blocks. Since that time, however, catch and market conditions have caused the operation to become more diversified. It is not unreasonable for Mr MacBaen to propose that the highly structured approach to certain situations (eg the broken knife; repair of the descaling machine) are indications that the organisation might benefit from a more flexible management structure in the future.

In his discussion with Mr Theobold, therefore, he suggested that an important factor in explaining the low output/employee might be the extremely rigid structure of the organisation. Mr MacBaen went on to suggest that the company might wish to examine ways of introducing a

96

more flexible approach and after further careful study, consider actions such as:

- Developing wider job specifications for employees, especially at the operative and supervisor levels.
- A major reduction of the formal procedures and associated paperwork which characterised the typical response to any situation requiring action between management levels and/or across departmental boundaries.
- Delegation of authority to lower levels in the operation.
- A wider span of control for supervisors and a reduction in the number of levels of management over the longer term.

Case: Queller Ltd

Queller is a company specialising in the production of processed seafood products for the catering industry. The organisational structure is based upon the traditional departmental divisions of responsibility (*Fig 9.1*).

During the last few years the company has faced increasing demands from their larger customers to move towards the formulation of individualised products more suited to their very different menu cycles. Although senior management are aware that meeting these type of requests does demand a more flexible approach to managing the organisation, the Managing Director is concerned to find that major delays are emerging during the development of any new, customer specific products. He had hoped that departments would be able to solve all problems by informal contact and liaison at all levels of the management hierarchy. It has become apparent, however, that all except the more urgent issues are being handled by executives referring problems upwards in their department and hoping that the senior managers will act to resolve inter-departmental requests for

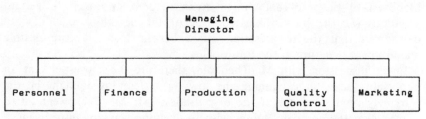

Fig 9.1   Current departmental organisational structure at Queller Ltd

action. This tendency is creating significant delays while senior managers confer, reach a decision and then communicate this decision back down to their respective subordinates.

The line-staff organisational structure

The structure shown in *Fig 9.1* is that of the standard grouping of tasks into functional areas of specialist activity. It is probably the commonest form of organisational structure and contains two types of managerial role – line management and staff management (hence 'line-staff' organisational structure). Within this type of structure the direction and control of a specific area of functional activity (*eg* processing of fish) is the responsibility of a line manager. Where this manager requires specialist outside assistance in working with the subordinates, this will be provided by a staff manager (*eg* disciplinary procedures being handled by the personnel department). This concept of differing managerial roles may be seen as containing aspects of both organic and mechanistic approaches to organisational management.

Line-staff organisational theory generally assumes that there are three types of staff management roles – personal, advisory, and control. The personal staff role (*eg* assistant to the Managing Director) involves assisting a manager but without having the authority to act on their behalf when they are unavailable. Advising staff (*eg* the personnel department) help line managers in areas where the latter require additional expertise. Control staff (*eg* quality control managers) are responsible for ensuring that line managers are meeting agreed objectives of performance.

The accountability of line managers is preserved by giving them the right to reject staff manager advice if they feel the advisor does not have an adequate understanding of the problems facing the line manager within his area of departmental responsibility. This right to reject advice is a potential source of conflict since staff managers with specific expertise do not like to have their capabilities challenged in this way.

Despite the risk of conflict between line and staff managers, this type of organisational structure is found to be extremely effective in organisations enjoying relatively stable market conditions. Consequently it is the most common form to be found in the fishing industry and is likely to remain completely acceptable in the foreseeable future. Nevertheless the structure should be carefully monitored to avoid the tendency to drift towards the more bureaucratic approach as illustrated by the Place Bay company. This is especially important because the fisheries and aquaculture industry now operates in a very uncertain business environment. There is need to accept a contingency-based approach to organisational design, *ie* senior managers should remain open-minded about introducing revisions in organisational structure if this should become a more effective approach to an unstable internal or external operating environment.

Functional departments cannot operate in isolation of other groups within an organisation. Although lines of authority flow upwards, effective co-ordination will depend upon both horizontal and vertical communication between managers during the execution of day-to-day tasks. Another potential hazard of the conventional line-staff structure is that horizontal communication is informal and is thus not capable of ensuring an immediate inter-departmental reaction to a major unforeseen problem which may suddenly confront the organisation.

Matrix management

The concept of matrix or project management originated within the aerospace industry as a response to coping with rapidly changing products and technology. The objective of the matrix approach is to overlay on the traditional vertical flow of authority, an additional horizontal line of authority and formalised responsibility as a way of improving the management of projects which require input from a number of departments. Managers involved in the project remain within their usual department, but work in a team made up of individuals from various areas of the company operations. The project manager will usually devote his whole time to the project, whereas individuals from the various departments will be involved on an 'as needed' basis, which may be as part-time or full-time contributors.

The project manager will be responsible for planning, performance attainment and control by working to unify the actions of all the individuals in the project team. However, the project manager has more responsibility than actual authority, so results are dependent upon persuasion, informal influence or appeals for co-operation rather than the more traditional coercive approach to enforcing managerial authority over subordinates.

The matrix concept is an extreme form of organic management, and there have been claims that it is superior to the more traditional line-staff structure. But matrix management is just as capable of creating problems as any other form of organisational structure. One of its commonest problems is that team members have trouble becoming used to the dual authority and reporting systems of the project manager and their respective departmental managers. Also when a team is disbanded upon completion of a project, members may find it difficult to settle back into their departments. Experience has shown, therefore, that the matrix approach functions best when the following conditions exist:

- Projects are to solve non-routine problems or react to non-standard market opportunities.
- The team members must be well trained and capable of contributing to a team-based environment.
- Projects should be those which require collaborative efforts from employees who have a wide range of differing skills and specialist knowledge.

Queller Ltd – revised management structure

The new strategy at Queller is to react to the specific market demands of individual key customers. To deliver the new products requires collaboration across virtually every department within the company. In order to overcome the delays inherent in the current organisational structure, one possible solution would be to create a matrix management structure (*Fig 9.2*).

Given the importance of meeting customer specifications for the new products, it would be advisable to draw the project managers from within

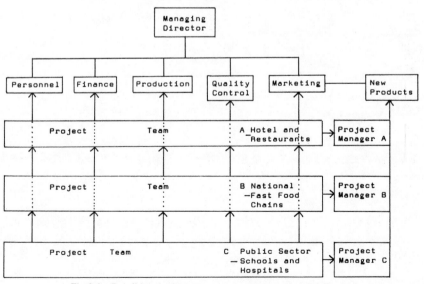

Fig 9.2 Possible matrix management structure for Queller Ltd

the marketing department because of their in-depth understanding of the customers' catering operations. These individuals might be given the role of specialist product managers responsible for three sub-sectors of the market – hotel/restaurants, public sector catering and national fast food chains – as shown in *Fig 9.2*. These three managers, although remaining as members of the marketing department, would have horizontal authority to call upon input from team members from other departments such as manufacturing or quality control. Contributions from these areas will tend to be on a part-time, not full-time basis.

10 The leadership role

Case: Kennedy Sales Ltd – Part I

Kennedy Sales is a large fish and seafood brokerage company which acts as sales agent for a number of producers from various parts of the world. They market products as both raw materials to companies involved in further processing, and direct to end users such as restaurants or wholesalers. The sales force has operated in the past across both the processor and end user sectors. However, because of the differing requirements of customers in these two sectors, the Board of Kennedy Sales has decided to reorganise the business into two separate selling operations. The 'Commodity Trading Group' will handle processors and the 'Commercial Sales Group' will cover the end user sector.

To establish the Commercial Sales Group, the company has hired Ray Keller who was previously a regional sales manager for a major seafood processor. Ray's success or failure in his new job will depend upon his capabilities as an effective leader.

A leadership model

Management researchers have for many years sought to find the answer to the question of 'what makes a good leader?'. Early studies focused on the 'trait approach' to leadership which proposed that leaders possess certain personality characteristics which explain their success. Unfortunately further research revealed good leaders who lacked many of the traits considered fundamental for success.

Discovery of leaders who apparently were exceptions to the trait theory caused management researchers to focus specifically on what successful leaders actually do at work. Again they encountered problems in establishing a unifying theory to fit every leadership type found to exist. To overcome this problem, management theorists have now accepted a contingency theory of leadership which hypothesises that the situation facing a manager will determine the most appropriate form of leadership behaviour.

A drawback to the contingency theory of leadership is that the practising manager seeking to understand how to be a better leader will conclude that all of this research is very interesting, but the proposed concepts are not easy to apply in real situations. Hence, before moving into any complex models of leadership, it is worth examining the functional aspects of leadership as proposed by John Adair. His theories are based on work carried out whilst a senior lecturer in military history and leadership training advisor in Britain in the '60s.

Dr Adair drew upon the extensive experience of the armed forces in the identification and development of military leaders. He combined these concepts with certain aspects of social psychology and proposed that there are three distinctive inter-related requirements of the individual placed in a leadership role. These he described as (a) definition of the group's task, (b) creation and co-ordination of the activities of the group involved in the task, and (c) development of the individuals who comprise the team. These requirements form the basis of the following types of 'need' and associated activities:

- *Task need* – which is the specification of the goal or objective towards which the team is working. Achievement of the objective will involve one or more tasks to be executed and a measurement of how the tasks are being executed is provided by comparing actual performance against the previously established objective.
- *Team need* – achievement of any objective will require the individuals in the group to work together in a co-ordinated and constructive fashion. The intent is to create synergy, *ie* the output of the team is greater than the sum of the efforts of the individuals involved.
- *Individual needs* – reflecting that each of the members of a group has their own areas of responsibility and deserve to know how they are performing as individuals. By receiving recognition of achievement, they may be motivated to perform even more effectively in the future.

The Adair model specifies that a leader must endeavour to fulfill all three areas of need – achieving the task, building the team and satisfying the needs of the individuals in the team. Should a leader concentrate too heavily on a single area (*eg* building team spirit but failing to clearly define the tasks of the group), it will be to the detriment of over-all performance required of the group by their organisation. In the example of over stressing team needs, the group would be working as a very cohesive group but fail to meet the organisation's expectations because the group has no clear idea of what standard of performance is required of them.

Achieving the task

To achieve the task assigned to a group, the leader must clearly understand the task and use this knowledge to define achievable objectives for the group. Given complete understanding of the assigned tasks and consequent objectives, the leader can develop a plan that will be used to brief the group on necessary actions. Development of a plan, although ultimately the responsibility of the leader, may be undertaken by this individual working in isolation or by involving subordinates in the planning process. Implementing this latter approach may be through allocating the entire task to the subordinates and asking for recommendations, or by working alongside the subordinates in a co-operative approach to planning.

Whoever is involved in the development of a plan, the leader must ensure that it contains details of the resources required, the time scale envisaged, and a definition of which member of the group will be allocated what aspects of the plan. In addition the leader will have to establish some form of control system which permits (a) an evaluation of actual results versus the plan and (b) identification of the causes for any variance from the plan which may occur.

Having defined a plan, the leader will need to make a detailed evaluation of whether resources specified will be available to the group, especially as they influence the workload of each of the team members. Should a capability gap become evident, the leader will have to consider how the productivity of the group might be altered (eg by further training or revising allocated responsibilities) or whether it will be necessary to recruit additional staff to handle the anticipated work-load.

Another issue the leader will have to examine is the capability of the team to perform their tasks in his absence. This may be possible through the definition of specified policies and procedures, but these might become inapplicable when faced with non-routine problems or an extended absence of the leader. The leader may therefore wish to consider nominating a deputy who has the ability and authority to act on the leader's behalf as required. The other benefit of training a deputy is that the leader is then perceived by his immediate superior as being more prepared for promotion, because there will be no discontinuity in group performance should the leader be promoted.

Finally a leader should review all objectives and plans with his superior to ensure that he has formal approval to proceed with executing the tasks required of his group.

Kennedy Sales – the task

At the point when the company divided the sales operation into two groups, the annual sales to end-users and wholesalers was as follows:

	£ million
Direct sales to hotel/restaurant chains	0.5
Sales to food service wholesalers	1.5
Direct sales to supermarkets	2.0
Sales to retail wholesalers	1.0

The company receives an average commission of 3% of sales from its principals, yielding a gross income of £150,000/annum. The assumption on operating expenses that Ray Keller has made for his Commercial Sales Group is:

	£'000
Salaries for managerial/sales staff	70
Operating expenses	30
Secretarial/administrative salaries	10
Office operation overheads	20
Total operating expenses	130

Ray decided that in the first year of operation it would be better to set a conservative sales forecast while he was beginning to gain a more detailed understanding of (a) market opportunity and (b) the strengths or weaknesses of his sales team. He assigned a sales objective of + 5% against prior year in all market sectors, but held expenditure at current levels. These objectives should result in an increase in net profit from £20,000 to £27,500.

Given the differences in product, promotional and operating goals of food service customers versus retail operations, Ray decided to allocate two salesmen to each sector. Each salesman would be responsible for specific accounts although they would be able to share a certain amount of work load during, for example, periods when special promotional events are being organised. Ray excluded himself from being responsible for any specific accounts, but made it clear he would be available to make key account calls with his salesmen if they felt his presence could be of benefit. Ray recognised that it was a priority to evaluate the capability of each member of the sales team as soon as possible. So in addition to setting quarterly sales objectives for each individual, he would need a control system which could identify the underlying cause of any variance from the specified sales objectives. This was to be achieved by introducing a weekly reporting system based upon information generated by the sales force describing the outcome of their customer calls. This information would be analysed by the secretarial/administrative staff at the office who would also generate additional information based on records of actual sales. These data could then express each salesman's performance relative to the average performance in the Commercial Sales Group through the use of the following ratios:

- Number of customers visited per week from the ratio of number of calls/week by the individual versus average calls per week within the sales force.
- The success in obtaining orders per week from the ratio of number of orders/week by the individual versus the average orders per week within the sales force.
- The size of order generated by accounts from the ratio of average order size/week for an individual versus the average order size by the sales force.
- The revenue generating capability of the salesman from the ratio of total sales generated per week by a salesman versus the average sales revenue generated by the sales force.

Meeting individual needs

A leader must never lose sight of the fact that a group is made up of individuals, all of whom require personal attention if objectives are to be met.

Motivation of each individual will only be achieved if each is gaining satisfaction from their work and also believes that they are making a worthwhile contribution. Additional requirements of individuals include receiving adequate recognition of their performance from the leader, and having control over the tasks which have been delegated to them. Most do not like to stand still in their career, so conditions should be created by the leader which allow subordinates to acquire new skills to prepare them for more responsibility in the future.

The role of the leader in ensuring the individual is effectively motivated will involve him in a careful analysis of each subordinate's potential versus actual role and achievement of goals versus assigned objectives. It will then be necessary for the leader to meet with each subordinate to mutually agree responsibilities and goals through which the subordinates can assess their own achievements. During the discussion of goals and responsibilities, the leader should ensure that the subordinate feels he has both the skills and resources to execute his assigned tasks.

In meeting the needs of the individuals in the group, the leader requires a very detailed knowledge of each subordinate's current skills relative to the tasks they will be expected to undertake.

Obtaining this understanding can be much simpler if the leader has up-to-date background information on every subordinate (eg personal details, education, previous employment history) as this will assist in the determination of what further actions might be required to optimise performance in the near future.

Developing the skills of each individual within the group is a very time consuming activity, but a very vital aspect of the leadership role. Hence the leader should endeavour to allocate a significant proportion of his time to working and meeting with each individual on a one-to-one basis. Employee development should never be approached as a task to be fitted in when possible (eg 'I've go a spare 30 minutes before lunch, maybe I'll talk to John and see how he is getting on'). Subordinates will react adversely to this casual type of behaviour and quite correctly conclude the manager is only paying lip service to this area of responsibility.

The ultimate objective of a leader is to reach the point where he has an accurate picture of each individual's aptitudes and attitudes within the work situation. Once he has this knowledge, it can be applied in providing each subordinate with a clear definition of their strengths and weaknesses while executing their assigned tasks. It also provides the basis for designing an appropriate programme for further developing the skills they will need to proceed smoothly along a career path in the organisation.

Kennedy Sales – individual needs

To illustrate the process of meeting individual needs, we can review the work of Ray Keller with one of the four salespersons, Graham Durant. Graham has been with Kennedy Sales for nine years and is considered to be a very experienced salesman. As his previous work was mainly in the retail sector, he was very pleased to accept Ray's suggestion that he should continue to concentrate on this type of customer following the reorganisation.

Ray had provisionally assumed that the two retail salesmen faced a similar sales potential in their territories. He therefore discussed with Graham his viewpoint on whether his accounts could generate annual sales of £1,575,000. Graham felt this might be possible but suggested he should break down his sales targets by key accounts to further validate the over-all sales target.

Ray's previous experience of sales force management had shown him that the most effective way of developing an effective sales team was to work in the field with subordinates on a regular basis. Hence he and Graham agreed on a specific day every month when Ray would accompany him. The defined structure of the day would be to have a couple of hours discussion in the office, followed by customer calls until 3 p.m., leaving two hours to review progress.

Ray was impressed with Graham's close working relationship with his accounts. They obviously considered him as an important source of guidance on fish availability and price trends. However, Ray observed that Graham rarely carried out any analysis of recent purchase patterns by a customer prior to making a call. Consequently discussions about future needs were often based upon Graham's own judgement, not informed recommendations.

Ray had previously discussed ways of building long term business in the retail sector with his sales team. There was common agreement that this could only be achieved by dropping the traditional practice of merely acting as order-takers, and beginning to work more closely with the accounts in all aspects of in-store merchandising of seafoods. While out with Graham, Ray also observed his apparent reluctance to discuss merchandising with the buyers unless they raised the issue.

Analysis of Graham's sales performance over the first quarter indicated sales at approximately 10% below budget and, furthermore, average order size per customer relative to the other retail salesman was also down by a similar proportion. Ray's opinion was that this reflected Graham's tendency to act only as an order-taker without paying sufficient attention to (a) understanding the customers' business when preparing for a sales call and (b) working with the customers to improve their in-store merchandising of Kennedy products.

Ray suspected that this reluctance reflected Graham's previous sales role and that he lacked experience in the new methods of managing retail accounts. At a quarterly review meeting with Graham he complimented him on his excellent relationship with his customers and then asked him if he had any areas of concern about his performance. Graham's response was that he knew he must do better but was not clear how to go about it. Ray then reviewed the areas of observed weakness and they agreed an action plan which included (a) arranging for Graham to attend a training programme in consumer goods selling, (b) having Ray make selected presentations to customers while Graham observed, (c) getting Graham to rehearse presentations before going out to customers and (d) using the two hours of the monthly joint selling day to evaluate strengths/weaknesses of Graham's handling of customers during the day.

Within two months, Graham's confidence in the new approach was reflected in both improved sales and also obtaining the first order from a major retail chain where Kennedy Sales had never been successful before.

Meeting group needs

A group represents the opinions and attitudes of the individuals who comprise its membership. It is vital for a leader to understand that his responsibilities include:

- Setting and maintaining the objectives for the group.
- Working with the group in achieving objectives.
- Developing skills within the group by using it as a forum for generating new ideas or approaches to resolving problems.
- Maintaining cohesion of the group and minimising dissent.
- Providing the group with opportunities for genuine consultation on any decisions which might affect their working conditions.

Kennedy Sales – the group task

Ray's approach to developing the salesmen to operate as a team was to schedule a full day meeting every month. This was held away from the office to avoid the distractions of telephone calls or other interruptions. The structure of the day soon evolved into a standard agenda of:

Ray: Briefing on over-all company performance and up-dating on any new corporate plans which might have impact on the Commercial Sales Group.

Staff: Individual presentations on recent activities and analysis of performance against objectives.

General: Discussion centring around a team-based approach to resolving problems and co-ordinating promotional programmes across accounts.

Issues: This part of the day was based upon input from individuals, reporting back on issues raised at the previous meeting and one or more items which Ray wished to cover.

In the first few meetings Ray noted a tendency of the salesmen to avoid raising issues which might be perceived as a personal weakness in their own performance, or to contribute to any debate about other people's performance. He realised that this reflected the previous business culture of Kennedy Sales where each salesman was left to their own devices and were not expected to become involved in matters outside their own job role.

After a couple of months, when Ray felt he had gained the group's confidence in his leadership, he introduced an outside training consultant to work with the group on developing their team membership skills. Once he felt the group were beginning to respond to this outside input, he specifically sought out agenda items about which he knew the group had strong opinions (*eg* delivery schedules to customers; responding to credit control department's requests for following up overdue accounts). Within six months, it was apparent that both in the meetings and in executing their normal selling responsibilities, there was a significant increase in the sales force working together to solve each others problems and to help each other out during periods of inequitable work loads.

11 Leadership style

Case: Kennedy Sales Ltd – Part II

At the end of the first year of the reorganised sales operation, Ray Keller was pleased that his team had exceeded the annual sales objective by 8%. It was therefore slightly irritating in a meeting with Tom Kennedy, the Managing Director, that – having briefly acknowledged this performance – Tom spent most of the time commenting on the Commercial Sales Group's handling of product ranges of three new principals.

He pointed out to Ray that these new principals were extremely disappointed with Kennedy Sales' handling of their business. In the first six months of the contract, sales had been minimal. Ray was already aware of the problem. He explained that he had been working with his sales team, emphasising the importance of the new principals and arranging training sessions on the most effective ways of presenting the products to potential customers. In his opinion this work would soon begin to pay dividends, especially as he had received indications that customers were favourably impressed with the new items.

Tom Kennedy shook his head and in his usual blunt fashion responded with, 'Your problem is that you place too much emphasis on this team approach to selling. You've only been here for a few months, whereas I've known your salesmen for years. Our new principals want results, not more training sessions. It's your job to make things happen in the Commercial Sales Group. So get back out there and if necessary start knocking heads. You can tell them from me, I know what's going on. They find it easier to get orders on existing items, and not spend time mentioning the new products. These new products are important to the long term future of Kennedy Sales and I'm not having them obstructed by your salesmen'. Without waiting for any response from Ray, he then went on to say, 'In the next few months we will probably be adding more new lines, so I need a Commercial Sales Group who can successfully obtain distribution on these types of product. As head of the group, I expect you to solve this problem; and while you're at it, I think you'd better do something about operating expenses. The monthly financial summary reports indicate that these are beginning to exceed budget'.

Leadership research

Very few people have the opportunity to be placed into a leadership position and even fewer, once given this role, are very successful. Earlier this century, researchers attempted to discover if there are certain characteristics or 'traits' exhibited by leaders which differentiated them from the rest of the population. Certain facets of personality such as a strong drive for responsibility, energetic approach to problem solving, capability to withstand stress and self-confidence were common to most successful leaders. Unfortunately, this research was found to be contradictory with common traits found in one study then being disproved in others.

This problem of validating earlier research studies has caused management theorists to move away from personalities and begin to look at what leaders do in organisations. One of the initiators of this latter approach was Ohio State University. They used the approach of comparing groups of leaders with groups of non-leaders and identified two aspects of the leadership role. One was the 'initiation of structure' which are the actions taken by leaders to get jobs completed through providing subordinates with plans, directions and controls. The second aspect 'consideration' was those actions where the leader recognises the human needs of subordinates and endeavours to optimise job satisfaction of employees.

Other research groups have also examined leadership behaviour and made further contributions to this field of study (eg University of Michigan; Harvard University). There were differences in methodology, analysis and conclusions reached. Nevertheless fundamental agreement has emerged that there are two dimensions to the leadership role:

- *Task orientation*, which is the extent to which leaders direct their own and subordinates' efforts; characterised by actions such as initiating, organising and directing.
- *Relationship orientation*, which describes the extent to which leaders involve themselves in the inter-personal relationships of the work environment; characterised by listening, trusting and encouraging.

Management style

The relationship between the task and relationship orientation can be illustrated through some form of management grid (*Fig 11.1*). The axes describe the relative importance given to the two forms of orientation and typically attempt to classify managers into exhibiting one of four possible types of style.

There is a certain degree of variation between management theorists on the terminology used to specify various styles of leadership. The one

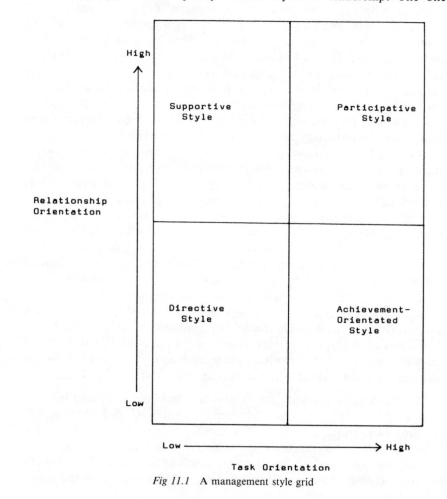

Fig 11.1 A management style grid

proposed by Robert House in the early '70s provides a very adequate example of the concept. In his model the styles are as follows:

- *Directive style* which is common among leaders who are very concerned about correcting deviations from established procedures and policies. Their perspective is to use past experience as the basis for carefully specifying what are deemed appropriate actions by subordinates. The approach enables subordinates to know exactly what is required of them and stresses the importance of logic or rationality in the approach to the problem/solution process.
- *Achievement-orientated style* which is prevalent among leaders who consider that achieving results is more important than any other objective. They expect subordinates to perform at their highest possible level of productivity and are confident that their subordinates willingly accept any new challenge. This type of leader relies heavily on their knowledge of the technical aspects of the job. Their approach to leadership is usually based on an overt capability to be able to do any job better than subordinates. To gain their support, this type of leader will rely heavily on clearly defined authority over the subordinates.
- *Supportive style* is characterised by leaders who are deeply concerned about the wellbeing of their subordinates. They have a genuine commitment to treating subordinates with openness and equality, always seeking new opportunities to improve the group's working environment. At all times, the objective of this type of leader is to achieve harmony between themselves and subordinates. Emphasis is on ensuring that subordinates are well informed and the most interesting tasks are delegated to the team.
- *Participative style* is a description of leaders who utilise the skills and capabilities of subordinates in reaching the best possible level of performance. Similar to the supportive style, development of a group-based approach to the problem/solution process is encouraged but not to the detriment of the quality of the decision. If participative leaders feel an issue cannot be delegated without risking an inadequate performance, this type of leader will carry out the task themselves.

Evaluating style

The usual approach in evaluating a leader's style is to have him complete a detailed survey which probes his likely behaviour during various situations that may be encountered in the organisation. For example, a manager might be asked to consider the following alternative responses (a) through (d) in the situation of 'setting objectives for yourself and subordinates':

(a) Goals should be clearly defined and should push everybody to improve their performance.
(b) Goals should be set which recognise subordinates' strengths and weaknesses.
(c) Goals should be set which develop team capability to maximise performance.
(d) Goals should be set on the basis of previous experience indicating what is realistically attainable.

The respondent is asked to locate responses (a) through (d) on a rating scale which describes the degree of similarity relative to the manager's usual behaviour.

Most similar	10	9	8	7	6	5	4	3	2	1	Least similar

Another question to be rated in the same way might be to consider (a) through (d) in controlling performance of subordinates:

(a) Ask for suggestions for possible improvements without introducing any drastic changes.
(b) Frequently review subordinates' progress against individual targets and determine capability.
(c) Use questions and dialogue to assess group performance.
(d) Utilise rigorous group-based meetings to review progress.

The four types of leaders will tend to allocate high scores (*ie* most similar to usual behaviour) as follows:

	Question 1	Question 2
Directive style	(d)	(a)
Achievement-orientated style	(a)	(b)
Supportive style	(b)	(c)
Participative style	(c)	(d)

Completion of a survey across numerous areas of activity will give the manager insight into which is their usual management style and an indication of the relative importance which they place on the four styles in the work environment. The differences in score between the first and second style favoured by the manager will indicate possible variations in style depending upon circumstances. A small difference would indicate a tendency to switch between styles (*eg* from participative to achievement-orientated) quite readily. A large difference could indicate a strong reliance on a single, predominant style.

The important fact to be stressed for any manager involved in a style assessment is that there is no style which is best. Some of the early researchers in this area did tend to emphasise one style (usually participative) for everybody. As training experts have gained more experience, it is now also recognised that individual flexibility in style is quite acceptable. Hence the objective of style analysis should be (a) to make managers aware of the implications of certain styles when working with subordinates and (b) that style can be modified to fit differing circumstances.

Risks of style rigidity

The research on leadership style has also been useful in identifying the potential risk in the excessive adherence to a single style. For example managers who have a directive leadership style may face difficulties when a non-routine problem is encountered. The manager will tend to fall back on previous experience and policies which may be inappropriate in a new situation. Subordinates may also interpret excessive adherence to this style as evidence that the leader is disinterested in their accomplishments. In an extreme situation, the subordinates may conclude that their role is that of automatons who are expected to operate a system which places reliance on adhering to rules, not attempting to improve performance.

The achievement-orientated style can easily deteriorate into complete autocracy where subordinates are ordered to complete tasks under impossible time constraints and with inadequate briefing. Their failure to complete the tasks will be met with threats of demotion or dismissal. At the extreme, the achievement-orientated leaders can actually misuse the power and authority vested in them by their superiors.

The supportive style which focusses on concern about group cohesiveness, has the potential to create dissatisfaction among subordinates. Individuals who would like a certain degree of self-responsibility may find that the leader is so interested in being involved in their tasks, there is no opportunity to act independently. His reaction to their dissatisfaction is to interpret the subordinate as lacking in group skills. Further stress is then created as the leader attempts to reintegrate the subordinate into the group.

In the extreme situation, important corporate objectives can be rejected as being incompatible with group needs.

The participative style offers two possible risks when applied in the extreme. The leader may become excessively supportive, resulting in the reduced independence of subordinates and the reduction of their motivation to optimise their performance. Alternatively, his desire to reach a decision may cause him to overrule the group opinion on an issue, and this immediately destroys the team relationship which he had apparently been striving to establish.

Style and subordinates

Having established the basic parameters of leadership style, further research has been carried out to examine the relationship between work tasks facing subordinates and the effectiveness of a specific style. These studies have shown that in meeting the leadership criterion of maximising the quality of any decision, one may violate the second criterion of reaching a decision that optimises the level of positive motivation among subordinates to implement the decision.

As long as the quality of a decision is not impaired, the leader may wish to consider modifying this style to fulfil objectives such as motivating subordinates, gaining wider acceptance as a capable leader or optimising subordinates' performance. Where the manager is considering a style modification, he should apply the general motivation concept that subordinates tend to be more satisfied with their job when they feel their efforts will lead to something they value. Hence the role of the leader is to utilise a style most likely to stimulate performance among subordinates. In proposing the concept that managers may wish to modify their style in relation to the characteristics of their subordinates, it is necessary to point out that there have been some contradictory findings in this area of leadership theory. The practising manager may therefore wish to treat the concept with a certain degree of caution.

When subordinates are faced with an ambiguous situation, they will prefer a directive style which uses procedures and rules to define their tasks (eg the installation of a new computer system in the accounting department). However, as the subordinates gain more experience of a new task, they may feel an excessively directive leadership style is too restricting and will respond better to a style which allows a greater degree of independence in making decisions. A move away from a directive to a more supportive style will also be effective when subordinates are facing a very stressful or frustrating set of tasks (*eg* staff in a personnel department involved in administering a redundancy scheme).

It is less clear which conditions motivate subordinates to respond better to an achievement-orientated style. Research has shown that white-collar employees involved in ambiguous, non-standard tasks appear to perform better under managers utilising this style. What is less certain, however, is whether these subordinates would react in the same way when working for a manager who has a participative leadership style.

For the practising manager, therefore, his choice of style should where possible reflect the behaviour patterns of subordinates. The most important issue to be kept in mind is for a manager to be consistent. Subordinates who have become used to a certain style may react adversely if the manager suddenly begins to use a new leadership style.

Style and the work environment

The activity of all subordinates is influenced to a varying degree by the conditions under which they work. Managers should recognise that differing environments may influence a subordinate's willingness to respond to a specific leadership style.

Under conditions where subordinates are involved in cerebral activities and the work method is clearly definable by procedures, then the directive style is very appropriate (*eg* carrying out analysis of variances in production costs). However, where subordinates are able to make skill-based decisions, have autonomy in selecting methodology, and can be creative, a supportive style is more appropriate (*eg* staff designing new training schemes for production operatives).

Where subordinates have insufficient knowledge or are likely to encounter unanticipated events needing corrective action by the manager, new instructions are frequently needed and performance is measurable, this type of work environment would favour an achievement-orientated style (*eg* production operatives being introduced to new equipment). However where subordinate interaction is necessary, effectiveness is influenced by interactions across the group and there may be a number of alternative solutions, a participative style would be more appropriate (*eg* a sales team presenting a new product to customers).

A manager must, however, be aware of the limitations placed on style by the organisational philosophy of the company. This philosophy is usually defined by senior executives and they will usually expect the philosophy to be reflected by managers at all levels in the company. So if company ethos is biased towards a very autocratic, structured form of operations, a junior manager who attempts to use a more supportive or participative approach can expect to encounter resistance.

This effect is also apparent when a company has operated under a certain style for many years and this style is now seen as the only acceptable approach. A new manager is unlikely to be able to introduce a new style unless the company is encountering major operational problems.

The choice of an appropriate style

The outcome of research on style and the widespread practice of including style assessment in training programmes, has in many cases left the practising manager in a dilemma as to how best to fulfill the leadership role. In some cases it may have been better not to have raised the issue in the first place, because all it has done is to confuse executives and reduce their own confidence in their abilities.

In recognition of this situation, a number of management experts have attempted to introduce some rationality which can assist the practising manager. One such concept has been proposed by Robert Tannenbaum, who suggests there is a range or 'continuum' of possible leadership styles depending upon (a) the necessity of the leader to retain control to obtain a desired result and (b) the degree of freedom that can be permitted among subordinates.

Where the leader has to retain absolute control, he will identify the problem, consider the alternatives, select one and inform the subordinates of his decision. In this situation he believes that no participation from subordinates can be accepted and if necessary, coercion may be utilised to ensure that his decision is implemented.

Where the leader feels that less authority or control is permissible, he will endeavour to 'sell' the decision to the subordinates. He recognises that there may be resistance and seeks to reduce this by persuading the subordinates of the benefits of the decision. When a greater degree of freedom can be tolerated, the leader will reach a decision and then review it and his analysis with his subordinates. They will be permitted to have an input, but nevertheless he makes it clear that he reserves the right to reject the advice of subordinates.

A more participative approach is where the leader defines the problem to the group, involves them in the consideration of alternatives and gives them the right to be involved in the decision. In the most extreme situations, the manager will leave the whole problem/solution process with the group and will accept their decision, even if he feels it is incorrect.

Kennedy Sales – flexible management style

Ray Keller tends to use a participative leadership style whenever possible. It is apparent that this is not totally to the liking of Tom Kennedy who feels a more achievement-orientated approach would be more appropriate.

It is apparent that Ray's style has failed to motivate his sales team to introduce the new products successfully. In applying the continuum style concept, therefore, this is a situation where Ray will have to retain complete control in reaching and implementing his decision.

Having carefully reviewed the situation, Ray believes that some form of coercion will be necessary in motivating his salesmen to reduce emphasis on selling existing products and directing more resources to the new items. The structure of the sales force remuneration is a basic salary plus bonus. The bonus in the first year was based on actual performance versus overall sales targets. Ray has therefore decided that in Year 2, 50% of the bonus will be reallocated to rewarding the sales force for the level of achievement in obtaining distribution of the new items in their accounts. He announced this decision at the monthly review meeting and made it quite clear that (a) his decision was irreversible and (b) reflected a failure by the sales force to react to his more participative requests for action on the new products.

On the issue of rising operating expenses, Ray felt that this could be dealt with in a more participative way. He therefore outlined the problem to the sales team and asked them to develop recommendations on alternative potential solutions. Even in this case, however, Ray indicated that the matter was of significant importance and that he retained the right to reject any solution which he felt was inappropriate.

12 Managing strategic change

Case: Bennett Corporation

Bennett Corporation is a regional shrimp processor producing breaded and peeled and de-veined (P&D) shrimp. The company also operates a commodity trading department which sells 'green headless' (*ie* shell-on) shrimp. The organisational structure is a standard functional type as shown in *Fig 12.1*, with each department headed by a Vice President.

Three years ago, the financial performance by product sector was as follows:

		£ millions		
	Breaded shrimp	P&D	Green headless	Total
Sales	10	15	15	40
Cost of goods	7	13.5	14.5	35
Gross profit	3	1.5	0.5	5
Operating expense	2	0.75	0.25	3
Net profit	1	0.75	0.25	2

Net assets 10   ROI 20%

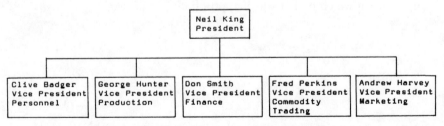

Fig 12.1 Organisational structure of Bennett Corporation

At that time, the Vice Presidents of Marketing and Commodity Trading developed a business plan designed to double the company's sales within ten years. The plan was approved by the Board of the company for immediate implementation.

The current financial situation is as follows:

	£ millions			
	Breaded shrimp	P&D	Green headless	Total
Sales	12	20	20	52
Cost of goods	8.7	18.2	19.4	46.3
Gross profit	3.3	1.8	0.6	5.7
Operating expense	2.0	0.9	0.3	3.2
Net profit	1.3	0.9	0.3	2.5

Net assets 13   ROI 19.2%

Mr Don Smith, Vice President Finance, has only been with the company for six months. He recently felt it useful to analyse financial performance over the last three years. His conclusion was that although absolute sales and net profitability have both increased, insufficient recognition had been given by senior management to the fact that (a) net profit as a percentage of sales has declined (from 5.0 to 4.8%), and (b) return on investment (ROI) has fallen (from 20 to 19.2%). The main cause of ROI reduction is an increase in net assets through the expansion by £2.5 million of the investment in green shrimp inventories.

Preliminary discussions with the Vice President of Marketing and Commodity Trading revealed that they saw little reason for concern. In fact their perspective was that the latest financial results indicated the company was well on the path to doubling in size within ten years.

Don Smith rapidly realised that his viewpoint was very much a minority opinion. He felt, however, that if the company's financial objectives were not modified, there would be a further erosion in net profitability and ROI. Under these circumstances, he was certain that Bennett Corporation's bankers would become concerned about the organisation's ability to cover future interest charges and loan repayments.

The need for change

Research on companies which fail has shown that such outcomes could usually have been avoided if senior management had accepted the need for a revision in strategy in response to changing business conditions. In theory a change in strategy should occur because top management have identified a gap between planned and actual performance. The problem is that managers seem to exhibit a sense of inertia, burying their heads in the sand and not being prepared to question variances in performance until it is too late. Even where evidence of growing problems are available, managers tend to ignore such information if it does not fit in with their own preconceptions. Furthermore once a crisis occurs, managers will often fall back on strategies which have been successful in the past, even though it is obvious that these are totally unsuitable for the new circumstances facing the organisation.

Strategic blindness seems most prevalent in companies which have been major forces in an industry in the past. Executives persuade themselves that any problem is only temporary in nature and can be overcome by a small increase in expenditure or investment. Unfortunately all that will occur in these situations is that when the company does eventually fail, the size of the losses facing creditors is even larger.

The executive who advocates a change in strategy has a long and difficult battle to fight. He will have to persuade others to revise their attitudes even before the topic of changing strategies can be accepted as a legitimate topic of conversation. Even if other managers can be persuaded to re-examine their beliefs, there will be further battles to be fought in gaining agreement to a new strategy, obtaining adequate resources to support new business plans, and persuading employees at all levels in the organisation to modify working practices.

Defenders of existing strategies usually outnumber those who advocate change. Furthermore these defenders are unlikely to give in easily. At best they will be seen by others as having made a mistake and at worst could lose their jobs. Hence through both covert and/or overt actions, the defenders will use every possible weapon available to discredit or undermine the position of the seekers of change.

Understanding the defenders

To have any chance of overcoming resistance to change among the defenders of current strategies, the manager must understand what is the basis underlying the opposition's position. It is likely that these can be grouped into the following categories:

- *Rejectors of logic* – these individuals are unwilling to even consider that the current strategy might be invalid. They will argue from the perspective that there is insufficient evidence for any change at this point in time. Their viewpoint can be based upon a genuine difference of opinion in the interpretation of the evidence presented to support the suggestion that there is a need for change. Unfortunately to defend their position, they may be willing to reject indications of a developing problem on the grounds that any downturn in performance is a short-term aberration. Under these circumstances the company may have to be on the verge of collapse before the defenders are willing to entertain any doubts about the wisdom of current strategies.

- *Acceptors of logic* – these individuals are willing to support the proposition of a need for change, but cannot be convinced that the new strategy is appropriate. Rejection of the solution may be based upon an opinion that there is a better solution and/or the proposed solution will divert resources away from other activities which they believe are more important.

- *Rejectors on principle* – this type of rejection is to be found among individuals who evaluate any solution in relation to potential impact upon their future role in the organisation. This form of rejection is the one which has been most deeply researched during studies on the management of change.

It is natural for most people to react to any new situation by considering how they personally will be affected by the proposed change. Unless they immediately perceive that such change is to their benefit, the natural response is to reject the proposition because it could leave them worse off in the future. There are a number of possible interpretations of how they would be adversely affected by change. Possibly they see that their skills would be rendered obsolete. Alternatively they recognise that new skills will be required and they are uncertain of their abilities to acquire new skills. There may also be concern about being placed in a new group and having to develop new working relationships in the work force. Another common cause of concern is that their job security will be threatened by redundancy, transfer to a new location, or loss of authority because someone will be promoted over them.

Gaining support for a proposal of change

When a strategy has been in place for some years, a coalition of support will have evolved. Unless the group all question the validity of perpetuating the current strategy, an individual who holds a different opinion will have a difficult time gaining acceptance for the need for change. Under these circumstances, the manager who holds a minority opinion will have to identify mechanisms to persuade others to agree with his position. This process will have to continue until the manager has generated sufficient support to be able to form an alternative coalition which has the strength to overcome opposition.

Attracting individuals to a cause may be achieved through education. The manager provides information which will cause others to question their current viewpoint because they begin to accept the merits of a superior argument. When other managers perceive a risk associated with the new concept, it may be necessary to offer incentives to weaken their resistance. The incentives will usually involve defining the benefits to the individual should the new strategy be adopted (*eg* increased authority; prospects for promotion).

Once a reasonable level of support has been achieved, it is likely that a committee will be formed to assess the strengths and weaknesses of the proposed change. The creation of the committee will make it legitimate for managers to discuss openly the issues associated with the proposed change. It is likely, however, that if there is going to be continued opposition, the formation of the committee will be seen as the signal for open warfare by the defenders of the current strategy.

Bennett Corporation – strategic change

Don Smith felt that the prime objectives for the company should be to improve both profitability and ROI. His examination of the situation led him to conclude that this could be achieved by restricting plans for sales growth to the higher margin breaded shrimp products while concurrently cutting back sales on P&D and green headless shrimp. The emphasis in reducing sales of commodity products should be on the green headless shrimp because this would reduce the company's need to hold large inventories of raw material. The consequent shrinkage in shrimp holdings would reduce net assets, increase ROI and lower the company's borrowing requirements.

To gain acceptance for the new strategy Don would have to convince the President and the other Vice Presidents. Through his understanding of these individuals, it was possible to forecast their reaction to his proposals as follows:

– Neil King, the President, would probably accept the logic but reject the proposed solution as his earlier successes in the company were based upon trading green headless shrimp.

- Andrew Harvey, Vice President Marketing, would reject the logic of the proposal on the grounds that he assisted in the development of the company's existing business strategy.
- Fred Perkins, Vice President Commodity Trading, would reject the logic of the proposal because the revised plan would endanger his role and status within the organisation.
- George Hunter, Vice President Production, and Clive Badger, Vice President Personnel, would both accept the new strategy as long as they could avoid being in conflict with other members of the management team.

It is apparent that to obtain support for his position, Don will have to gain acceptance for his ideas from Neil King and Andrew Harvey. The latter's support was vital because only he had the expertise to validate the possibilities of expanding sales for breaded shrimp. Andrew had always indicated a willingness to accept concepts if they would strengthen his position in the company. Hence Don felt that an incentive approach of needing to expand the company marketing operation would be of appeal to Andrew.

Don arranged to have lunch with Andrew and, without revealing any concerns about profitability, raised the issue of the practicalities of expanding the breaded shrimp business. Andrew indicated that this was possible as long as additional financial resources were made available to support the hiring and training of more sales staff in the food service sector.

Neil King had always seemed very open-minded about other proposals which Don had made. Hence he felt that an education approach could be used to obtain the President's acceptance of the revised strategy. Don therefore submitted a confidential report which highlighted the profit implications of sustaining the current strategy and described the potential benefits of expanding sales for breaded shrimp. Neil's reaction was to request that Don form a small working party with Andrew Harvey and George Hunter. Neil considered that the latter's input would be vital because of the implications on plant capacity and/or capital expenditure to handle any additional output of processed shrimp.

Don was pleased to find that Neil was perceptive enough to recognise Fred Perkins' potential opposition. He therefore suggested that at this stage the working party should only examine the practicalities of expanding breaded shrimp sales and avoid any discussion of the second phase of the revised strategy involving de-emphasis of sales for P&D or green headless shrimp.

Resolving conflict

It is usually impossible to introduce change without creating conflict as managers seek to retain their current span of authority within the organisation. The conflict will not just occur during the planning phase, but will continue on as new policies are introduced that modify, alter or abolish existing working practices throughout the organisation. In order to minimise conflict, managers will have to use a variety of approaches if new strategies are to be successfully implemented.

One of the commonest approaches to conflict management is to act 'coercively', where employees are told to accept change and that any resistance could mean job losses or reduced promotional prospects. In some circumstances, the employees do not believe these threats and it may be necessary to actually fire a few individuals to persuade the workforce that the management is serious about introducing new working practices. The major risk of a coercive approach is that it will exacerbate employees' resentment to the company's revised policies. Nevertheless in situations where speed of action is essential (eg without change, the company will go bankrupt) and the changes are known to be unpopular, then coercion may be the best option to manage conflict.

Where time permits, however, it is clearly preferable to use a more participative approach. By involving employees in the change process at the earliest possible point, the greater is the likelihood that they will react positively. For the participative approach to be effective, management should provide detailed information on both the need for, and the logic behind, the proposed changes. But this process will only be effective where the manager has already established a relationship of trust with the audience to whom the information will be delivered.

In this approach, communication of any type is presented in a form which permits a two-way dialogue in which the receivers of the information are provided with an opportunity to ask questions or propose alternative solutions. When subordinates can clearly understand the reasons for rejecting any of their suggestions, it is more probable that they will accept the manager's original proposals. If any manager indicates an interest in receiving comments from subordinates, it is vital that the subordinates believe this to be a participative dialogue and not a manager just paying lip service to the concept of participative discussions.

The participative approach is unlikely to be productive in an atmosphere of widespread fear and anxiety (real or imagined) about a proposed change. Under these circumstances, the manager should attempt to be supportive and spend time understanding the underlying causes of the fears being expressed by subordinates. In many instances, these type of discussions can be used to introduce the issues of the new skills that will have to be acquired by subordinates before they can effectively execute their revised job roles.

Another mechanism for resolving conflict can be to offer incentives to reduce opposition. This method will usually involve negotiations during which the two parties are willing to modify their stance in return for additional benefits. An example would be an agreement to modify work practices in return for a bonus based on higher productivity. In those situations where subordinates are members of a union, this organisation will often provide a full time official to negotiate on behalf of the work force.

Managers sometimes use the mechanism of co-option to resolve conflict. This approach involves including an individual in the planning or implementation phase in return for their endorsement of the proposed change. The risk associated with this method is that the co-opted individual assumes that their opinion is actually important. At the point when they recognise the fallacy of this assumption, a common reaction is to covertly oppose the change process. Their ability to succeed in this action is aided by the fact that through co-option they have gained access to confidential information. Hence co-option, although popular, is not usually very effective. Where a manager is frequently seen to use this approach, subordinates become very wary of the manager and endeavour to minimise contact with him to avoid being drawn into the change process.

No manager can expect to identify a single approach that can be applied to every conflict resolution situation. It is therefore important that the manager examines the specific circumstances of any situation and then determines which is the most appropriate mechanism to be used. A continuum of choices are available ranging from coercion at one extreme through an extremely participative approach at the other. Where a manager faces massive resistance, then a participative style is more likely to succeed than acting dictatorially. This is especially true where the manager is completely dependent upon wholehearted co-operation of subordinates when implementing a proposed change in work practices.

Where the manager expects minimal resistance, or the change has little impact on the majority of the staff, coercion will prove quite feasible. When the risks are high or an immediate change necessary, then this is also a situation where coercion should be used. The approach is also mandatory where the initiator has little power over others. This latter situation may at first appear somewhat illogical, but only managers who have already established a strong level of authority over others are able to tolerate a participative approach when managing change.

Bennett Corporation – handling opposition

Don Smith already recognised that Fred Perkins represented the most likely source of opposition. The creation of the working party was certain to be the signal for open warfare over changing the company strategy. After some reflection on this fact, Don therefore raised his concerns with Neil King. He agreed that before the working party was formed, it would be necessary to review the situation with Fred. Neil's previous experience led him to conclude that a coercive approach would be required and that he, as President, should initiate a discussion with Fred. Furthermore, at this meeting, he would make it clear that the working party had presidential support for examining the concept of expanding breaded shrimp sales.

Don was then able to brief the other two members of the working party, knowing that for the moment Fred's opposition had been neutralised. Given the differing expertise of the working party, Don decided to use a very participative approach in the analysis of the proposed changes in strategy. Their deliberations over a four-week period validated Don's proposals and a formal presentation to the complete senior management was scheduled.

As expected, Fred Perkins was extremely critical and focussed upon the fact that Don's limited experience of the shrimp industry was the reason for reaching an incorrect conclusion about the future direction for the company. These comments were received with some sympathy by Clive Badger and Neil King and sensing victory, Fred suggested that the new strategy be shelved and possibly re-examined in a 'couple of years time'. At this point both George Hunter and Andrew Harvey made it clear that their involvement in the working party had caused them to completely accept the need for an immediate change in strategy. This very positive new coalition eventually caused everyone except Fred to adopt the new strategy of expanding the breaded shrimp operation with a concurrent reduction in commodity trading activities.

Implementing change

The planning phase of the change process usually only involves a small number of individuals within an organisation. Implementing change is typically the point when a much larger proportion of the work force are included. Prior to making any announcements, therefore, it is vital that managers carefully analyse the implications of any change and draw up detailed plans on what actions will be required at all levels of the organisation. Once this has been done, the managers should schedule meetings to brief subordinates on new roles and revisions in activity. Where it is intended to have a participative approach in the change process, sufficient time should be scheduled to permit completion of the discussions with subordinates.

Unless the organisation faces a crisis which mandates immediate revision in job roles, attempts should be made to extend the introduction of change over a time period sufficient to cope with any unforeseen problems which may develop. There are a number of approaches that can be used in the resolution of unexpected obstacles. One possibility, especially when new skills will have to be acquired, is to use a phased introduction geared to the pace of the training programmes to be used. Other approaches include pilot schemes involving only some of the work force or to have rehearsal periods prior to the actual change. When using these latter mechanisms, the manager should encourage subordinates to comment on their own performance and contribute alternative ideas which might improve efficiency.

When the organisation is concerned that very major problems may be encountered, it is not unusual to operate parallel systems when the existing and new work practices are operated side-by-side until the new method has been completely validated. This method, for example, is often used when companies move from manual to computerised control systems.

Bennett Corporation – implementing the new strategy

The adoption of the new strategy will have ramifications across the entire organisation. Broad-scale acceptance by all department heads would be unlikely if Don Smith was seen to have been responsible for the new approach. In this situation, it is imperative that the President, Neil King, is seen as the prime agent in introducing the new strategy.

Both the Marketing and Production Departments will have to produce very detailed plans which translate the strategy of expanding the breaded shrimp operation into action documents defining the responsibilities of staff in these two departments. Briefing sessions will then be necessary to ensure that everybody comprehends the implications of the new strategy. Initially there will be severe pressure on staff in these two departments until new employees are recruited or transferred from other departments to handle the heavier work load.

At the same time the President will have to cope with the negative reaction of de-emphasising the commodity trading operation, especially as the implications of the new strategy are that some staff will be transferred to other departments. A critical element in the transition will be the behaviour of Fred Perkins. If there is any sign that he is either actively or passively obstructing adoption of the new strategy, then the President — despite their long standing personal friendship — would probably have no other option but to request Fred's resignation.

Further reading

Adair J. *Action Centred Leadership*. Gower Press 1979
Child J. *Organisation: A Guide for Managers and Administrators*. Harper & Row 1977
Drucker P. *Management: Tasks, Responsibilities and Practices*. Harper & Row 1974
Hertzberg F. *Work and the Nature of Man*. World 1966
Humble J. W. *Management by Objectives in Action*. McGraw-Hill 1970
Maslow A. *Motivation and Personality*. Harper & Row 1954
Mintzberg H. *The Nature of Management Work*. Harper & Row 1973
Porter L. W. and Lawler E. E. *Management Attitudes and Performance*. Irwin-Doney 1969
Reddin W. T. *Managerial Effectiveness*. McGraw-Hill 1970
Stewart R. *Choices for the Manager*. McGraw-Hill 1982
Vroom V. *Work and Motivation*. J. Wiley & Sons 1964
Vroom V. and Yellon P. *Leadership and Decision Making*. University of Pittsburgh Press 1973
Zaltman A. D. and Duncan R. *Strategies for Planned Change*. Harper & Row 1977

Books published by
Fishing News Books Ltd

Free catalogue available on request

Advances in fish science and technology
Aquaculture practices in Taiwan
Aquaculture training manual
Aquatic weed control
Atlantic salmon: its future
Better angling with simple science
British freshwater fishes
Business management in fisheries and
  aquaculture
Cage aquaculture
Calculations for fishing gear designs
Commercial fishing methods
Control of fish quality
The crayfish
Culture of bivalve molluscs
Design of small fishing vessels
Developments in fisheries research in
  Scotland
Echo sounding and sonar for fishing
The edible crab and its fishery in British
  waters
Eel culture
Engineering, economics and fisheries
  management
European inland water fish: a multilingual
  catalogue
FAO catalogue of fishing gear designs
FAO catalogue of small scale fishing gear
Fibre ropes for fishing gear
Fish and shellfish farming in coastal waters
Fish catching methods of the world
Fisheries oceanography and ecology
Fisheries of Australia
Fisheries sonar
Fishermen's handbook
Fishery development experiences
Fishing boats and their equipment
Fishing boats of the world 1
Fishing boats of the world 2
Fishing boats of the world 3
The fishing cadet's handbook
Fishing ports and markets
Fishing with light
Freezing and irradiation of fish

Freshwater fisheries management
Glossary of UK fishing gear terms
Handbook of trout and salmon diseases
A history of marine fish culture in Europe
  and North America
How to make and set nets
Introduction to fishery by-products
The lemon sole
A living from lobsters
Making and managing a trout lake
Managerial effectiveness in fisheries and
  aquaculture
Marine fisheries ecosystem
Marine pollution and sea life
Marketing in fisheries and aquaculture
Mending of fishing nets
Modern deep sea trawling gear
More Scottish fishing craft and their work
Multilingual dictionary of fish and fish
  products
Navigation primer for fishermen
Netting materials for fishing gear
Ocean forum
Pair trawling and pair seining
Pelagic and semi-pelagic trawling gear
Penaeid shrimps — their biology and
  management
Planning of aquaculture development
Refrigeration on fishing vessels
Salmon and trout farming in Norway
Salmon farming handbook
Scallop and queen fisheries in the British
  Isles
Scallops and the diver-fisherman
Seine fishing
Squid jigging from small boats
Stability and trim of fishing vessels
Study of the sea
Textbook of fish culture
Training fishermen at sea
Trends in fish utilization
Trout farming handbook
Trout farming manual
Tuna fishing with pole and line